Current Hypotheses and Research Milestones in Alzheimer's Disease

Ricardo B. Maccioni · George Perry
Editors

Current Hypotheses and Research Milestones in Alzheimer's Disease

Tim

To my good friend
and collegue.

George Perry

4 December 2009
San Antonio

🐎 Springer

Editors

Ricardo B. Maccioni
Mind/Brain Program of the International
Center of Biomedicine and
University of Chile
Santiago, Chile
and
University of Miami School of Medicine
Miami, Florida
USA

George Perry
College of Sciences
University of Texas at San Antonio
San Antonio, Texas 78249-0661
USA

ISBN: 978-0-387-87994-9   e-ISBN: 978-0-387-87995-6
DOI: 10.1007/978-0-387-87995-6

Library of Congress Control Number: 2008942056

Printed on acid-free paper

springer.com

Preface

Almost four decades of increasingly faster and more intensive research on Alzheimer's disease (AD) have brought major advances in our understanding of its pathogenesis, improved tools for diagnosis, and strategies for its treatment. This research has shown us that exploring the factors leading to cognitive impairment is mired with elusive goals and unexpected leads toward the mechanisms responsible for this pathology. This complex, yet fruitful process has allowed us to contribute to building a solid foundation of knowledge in the neurosciences, to tackle the biological basis of AD, and related neurological disorders.

This book contains 20 seminal chapters by authors with different views that in a multidisciplinary approach focus on the main issues of neuroanatomical, neuropathological, neuropsychological, neurological, and molecular biological aspects of AD. The idea for this collection arose during the preparation of the proceedings for the International Summit Meeting "Current Hypotheses on Alzheimer's Disease" and was warmly received by the prestigious editorial house Springer Verlag. The book you now have the opportunity to read is the result of substantial efforts by many people, including the editors and their collaborators. The main goal has been to summarize in several chapters the contributions presented in an international gathering held November 22–25, 2007 at the Conference Town in Reñaca, Viña del Mar, Chile, attended by some of the leading scientists in the field. Presentations focused on milestones in research that have illuminated Alzheimer's investigations during the past decades and will continue at the forefront.

This prolific event was organized by the International Center for Biomedicine (ICC), a leading center of excellence established in Chile that has contributed to research for decades, fostering neuroscience research, advanced education, and international collaboration. For many years, the ICC has been at the forefront of AD research, organizing numerous symposia, international laboratory courses, and publications on key discoveries in the biomedical sciences.

AD is the most prevalent cause of dementia throughout the world, and the fourth cause of death in developed economies after cancer, cardiovascular disease, and stroke. AD afflicts over 5 million people in the United States, about 12% of those older than 65 years old, and at least half of those older than 85 years of age. Furthermore, considering that AD accounts for the largest number of cases of dementia, and that other conditions, including dementia with Lewy bodies, fronto-temporal dementia (FTD-17), and vascular dementia, also increase in occurrence with age, the steadily increasing prevalence of these pathologies in numerous aging societies all over the world constitutes one of the major medical and scientific challenges on a global scale. Unfortunately, we still do not have reliable and definitive biomarkers for AD, and a certain diagnosis is still possible only postmortem. Many hypotheses have been proposed on the biological basis for this disease, and even more are the subjects of discussion on the World Wide Web, a veritable "factory" of AD hypotheses. Addressing this situation, the purpose of the meeting in Chile was to bring together a representative group of those who have developed the most promising and challenging hypotheses to date, to attempt to integrate these ideas into a common theory of AD.

Dialogue is beyond baptists, tauists, or agnostics. AD is developing as a pleotropic disease closely linked to the aging processes and involving cytoskeletal proteins, kinases, radicals, metals, genetics, inflammation, viruses, nucleation, and amyloid α, with all playing a critical and complex role. Reductionism has not reduced AD any more than for other age-related degenerative diseases. Changing perspectives mark the beginning of a second century of AD.

The experience of preparing this book has been very rewarding, and we are grateful to all the contributors who submitted their chapters and were collaborative in the exhaustive reviewing process at the editorial level. We thank especially Bethany Kumar for her invaluable support in coordinating efforts among editors and authors, and Nancy Rawls for assisting with the editing process. We also thank numerous attendants, students, postdoctoral fellows, and health professionals who participated in the event and who contributed helpful viewpoints, criticisms, and incisive questions that enriched the contributions to this book.

We hope that this book will mark the beginning of a tradition of developing exciting meetings on the theoretical foundations guiding research into very challenging dilemmas in biomedicine, and thus contribute to solving major public health problems that afflict mankind.

Santiago and San Antonio, 2008   Ricardo B. Maccioni
                                                              George Perry

Contents

General Aspects of AD Pathogenesis

Biomarkers and AD

Cognitive Neurology in AD

Amyloid–β: General Aspects

Role of Aβ Degrading Enzymes in Synaptic Plasticity and Neurogenesis in Alzheimer's Disease

Leslie Crews, Brian Spencer, and Eliezer Masliah

Abstract Alzheimer's disease (AD) is a leading cause of dementia in the aging population. This progressive neurodegenerative disorder is characterized neuropathologically by the presence of plaques composed of amyloid-β (Aβ) peptides and neurofibrillary tangles containing phosphorylated tau; however, neurodegeneration in AD probably initiates with damage to the synaptic terminal. In addition to the neurodegenerative features of AD, the pathological process in this disorder is accompanied by significant alterations in adult neurogenesis in the hippocampus. Although the precise mechanisms leading to neurodegeneration and neurogenic alterations in AD are not completely understood, several lines of investigation indicate that enzymatic cleavage of Aβ may play an important role in preventing or reversing the neurodegenerative process in AD. Neprilysin (NEP) is one such Aβ-degrading enzyme, and in addition to its Aβ-dependent effects, NEP is capable of promoting neurogenesis and synaptic remodeling. Since this endopeptidase is capable of cleaving a wide range of neuropeptides with neurotrophic activity, its trophic action in the CNS is probably related to enzymatic processing of substrates such as neuropeptide Y (NPY) that results in the generation of neuroactive products. Thus, NEP represents a unique example of a proteolytic enzyme with dual action: degradation (Aβ) and processing (NPY); both actions are neuroprotective. Therefore, understanding the effects of NEP on NPY and other neuropeptides might provide new information about the neuroprotective mechanisms of NEP in the mature CNS and in animal models of AD.

Keywords Synapse loss, Neurodegeneration, Neurogenesis, APP, Neprilysin

L. Crews, B. Spencer, and E. Masliah (✉)
Department of Neurosciences University of California, San Diego, La Jolla, CA 92093
e-mail: emasliah@UCSD.edu

R.B. Maccioni and G. Perry (eds.) *Current Hypotheses and Research Milestones in Alzheimer's Disease*. DOI: 10.1007/978-0-387-87995-6_1,
© Springer Science + Business Media, LLC 2009

3

1 Introduction

Alzheimer's disease (AD) continues to be the leading cause of dementia in the aging population [1]. Over 5 million people live with this devastating neurological condition and it is estimated that the USA will experience an average 50% increase in patients with AD by the year 2025 [2]. AD is a progressive neurodegenerative disorder that specifically damages limbic structures, the association neocortical pathways [3–6], and the cholinergic system [7, 8]. Although the key neuropathological diagnostic features of AD are the presence of plaques – composed of amyloid-β (Aβ) peptides and the neurofibrillary tangles containing phosphorylated-τ [9], the neurodegenerative process in AD probably initiates with damage to the synaptic terminals [10, 11]. It has been postulated that the early synaptic pathology leads to axonal abnormalities [12], spine [13], and dendritic atrophy [14] and eventually neuronal loss [11, 15].

Although the precise mechanisms leading to neurodegeneration in AD are not completely understood, several lines of investigation indicate that alterations in the amyloid-β protein precursor (AβPP), resulting in the accumulation of Aβ and AβPP C-terminal products, might play a key role in the pathogenesis of AD [16–20] (Fig. 1a). Several products are derived from AβPP through alternative proteolytic cleavage pathways, and enormous progress has recently been made in identifying the enzymes involved [21–25] (Fig. 1b). Therefore, disruption of the mechanisms involved in modulating synaptic plasticity might be responsible for the characteristic cognitive deficits in AD patients and as such represent an important target for treatment development.

2 Alterations in AβPP Processing, Synaptic Plasticity, and Neurogenesis in AD

The most significant correlate to the severity of the cognitive impairment in AD is the loss of synapses in the frontal cortex and limbic system [10, 11, 26, 27]. The pathogenic process in AD involves alterations in synaptic plasticity that includes alterations in

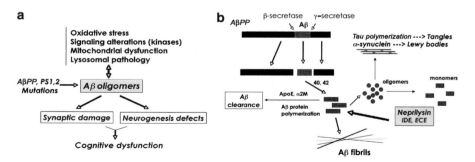

Fig. 1 Mechanisms of Aβ toxicity and clearance. (**a**) Accumulation of Aβ oligomers might be involved in promoting synapse damage and neurogenesis defects. (**b**) Aβ-degrading enzymes such as neprilysin (*NEP*), insulin-degrading enzyme (*IDE*), and endothelin-converting enzyme (*ECE*) play a central role in Aβ clearance (*See Color Plates*)

formation of synaptic contacts, changes in spine morphology, and abnormal area of synaptic contact [28]. However, other cellular mechanisms necessary to maintain synaptic plasticity might also be affected in AD [29–31]. Recent studies indicate that neurogenesis in the mature brain plays an important role maintaining synaptic plasticity and memory formation in the hippocampus [32] (Fig. 2).

In the adult nervous system, motor activity and environmental enrichment (EE) have been shown to stimulate neurogenesis in the hippocampal dentate gyrus (DG) [32, 33]. Interestingly, studies in human brains [34] and transgenic (tg) animal models have demonstrated significant alterations in the process of adult neurogenesis in the hippocampus in AD [35–39]. The deficient neurogenesis in the subgranular zone (SGZ) of the DG found in our AβPP tg mice [40] (Fig. 2) is consistent with studies in other lines of AβPP tg mice and other models of AD that have shown decreased markers of neurogenesis, such as bromo-deoxyuridine (BrdU)+ and doublecortin (DCX)-positive cells, with an increase in the expression of markers of apoptosis [39, 41–43]. Although a different study reported increased neurogenesis in the PDAβPP model [38], a more recent and comprehensive analysis showed that while in the molecular layer (ML) of the DG there is an increased number of NPC, in the SGZ markers of neurogenesis are decreased, indicating that in PDAβPP animals, there is altered migration and increased apoptosis of NPC that contributes to the deficits in neurogenesis [37].

Thus, alterations in synaptic plasticity in AD might not only involve direct damage to the synapses but also interfere with adult neurogenesis (Fig. 1a). The mechanisms of synaptic pathology in AD are the subject of intense investigation. Studies in experimental models of AD and in human brain support the notion that aggregation of Aβ, resulting in the formation of toxic oligomers rather than fibrils, might be ultimately responsible for the synaptic damage that leads to cognitive dysfunction in patients with AD [44–46] (Fig. 1a). Supporting this notion, it has been shown

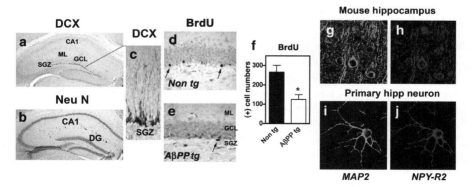

Fig. 2 Neurogenesis in the hippocampus in AβPP tg mice and neuropeptide (*NPY*)-R expression. (**a–c**) Doublecortin (*DCX*)-positive neuronal precursor cells (*NPC*) in the hippocampus subgranular zone (*SGZ*). (**d and e**) In the mThy1-AβPP tg mice, the numbers of NPC in the SGZ are reduced. (**g–j**) NPY-R2 is colocalized in the neuronal marker MAP2 in the mature hippocampus and in primary hippocampal cultures ($n = 12$ mice per group) (*See Color Plates*)

that Aβ oligomers reduce synaptic transmission and dendritic spine movement [14, 44, 47], and interfere with axoplasmic flow and activate signaling pathways that might lead to synaptic dysfunction, τ-hyperphosphorylation, and cell death. Moreover, a dodecameric Aβ complex denominated *56 has been recently characterized [48] in brains from AβPP tg mice and shown to contribute to the behavioral alterations in these animals. The differential effects of this and other toxic oligomeric arrays of Aβ in mature and developing neurons and synapses await further investigation.

The accumulation of Aβ in the CNS and the formation of toxic oligomers most likely depend on the rate of Aβ aggregation, synthesis, and clearance (Fig. 1b). Although most effort has been concentrated at elucidating the mechanisms of Aβ production and aggregation [22–24, 49, 50], less is known about the mechanisms of Aβ clearance. This is important because while most familial forms of AD might result from mutations that affect the rate of Aβ synthesis and aggregation, sporadic AD might be the result of alterations in Aβ clearance (Fig. 1b). Pathways involved in Aβ clearance include binding to LDLR-related protein (LRP) ligands such as apolipoprotein E (ApoE) [51, 52], lysosomal degradation [53, 54] and cleavage by proteolytic enzymes such as neprilysin (NEP), insulin-degrading enzyme (IDE), angiotensin-converting enzyme (ACE), endothelin-converting enzyme (ECE), and matrix metalloproteinase-9 (MMP9) [55–61] (Fig. 1b).

We utilized lentiviral vectors to investigate the ability in vivo of some of these Aβ-degrading enzymes at clearing amyloid and ameliorating the neurodegenerative pathology in tg mice [62, 63]. We found that NEP was capable of reducing plaque load, soluble Aβ levels, synaptic pathology, and behavioral deficits. This is consistent with previous studies using tg mice or other viral vectors [63–65], and supports a role of NEP in the treatment and pathogenesis of AD.

3   The Role of Aβ-Degrading Enzymes in the Clearance of Aβ Aggregates and the Pathogenesis of AD

Considerable progress has been made toward identifying endopeptidases, which directly degrade Aβ and play an important role in the homeostatic control of this peptide. Among them, NEP (also known as CD10 or EC 3.4.24.11) – a zinc metalloendopeptidase [66] – has been identified as a critical Aβ-degrading enzyme in the brain [56, 57] (Fig. 1b). Other NEP family members, for example ECE, may contribute to amyloid catabolism and may also play a role in neuroprotection. Another metalloproteinase, IDE, has also been advocated as an amyloid-degrading enzyme [59] and may contribute more generally to metabolism of amyloid-forming peptides. Other candidate enzymes proposed include ACE, some matrix metalloproteinases, plasmin, and, indirectly, thimet oligopeptidase (endopeptidase-24.15) [60, 67].

NEP is capable of degrading both Aβ monomers and oligomers (but not fibrils) [68] (Fig. 1b). NEP levels are reduced in the brains of AD patients [69–73] and a potential genetic linkage is currently being investigated [61, 74–76]. Further supporting a central role for NEP in AD, previous studies have shown that chemical inhibition of NEP with thiorphan or crosses between AβPP tg and NEP-deficient mice results in rapid accumulation of Aβ and pathological deposition in rodents [57, 77]. Moreover, we and other groups have shown that overexpression of NEP by gene transfer with viral vectors [63–65], neuronal promoters [59], or induction [78] resulted in a reduction in amyloid pathology [76, 79–81]. NEP concentrates at the synaptic terminals [64] where it is most effective at reducing Aβ oligomers [68]. Although considerable effort has been placed at investigating the effects of NEP on Aβ levels, less is known about the trophic effects of NEP, its effects on other AβPP products and substrates, and to what extent this might play a role in the mechanisms of synaptic plasticity, neuroprotection, and neurogenesis.

Remarkably, under basal conditions, levels of NEP expression might be regulated by neuronal activity similar to what has been described for neurogenesis. For example, gene array studies in AβPP tg mice exposed to increased physical activity have shown that reduction in the AD-like neuropathology might be associated with increased NEP expression [82]. Moreover, our recent studies have shown that in addition to the effects on Aβ levels, NEP is capable of promoting neurogenesis and synaptic remodeling [62, 83, 84]. Although it is clear that physical activity promotes neurogenesis and learning in the hippocampus [85], it is still controversial what is the effect of physical activity and intellectual EE on amyloid deposition. While some studies have shown a reduction [82, 86], others have shown either no changes [87] or an increase [88]. Differences might be related to the characteristics of the EE and physical activity, the gender, and the AβPP tg line. A recent study suggests that combined physical activity and intellectual EE (referred as "complete enrichment") is necessary to reduce Aβ load and increase synapses in AβPP tg mice [89].

Taken together, these data suggest that in addition to the Aβ-dependent effects, NEP might exert a trophic action in the CNS via Aβ-independent pathways. In addition to Aβ, the endopeptidase NEP is capable of cleaving a wide range of neuropeptides with neurotrophic activity, including SOM, SP, ENK, and NPY [67], and might regulate the activity of growth factors. Studies by Saito et al. [55] have shown that SOM and NEP coregulate their expression and we have recently shown that NPY fragments resulting from NEP proteolysis might be neuroprotective [83] (Fig. 3). NPY, a 36-amino acid long molecule, is the most abundant neuropeptide in CNS and NEP cleaves NPY at the C-terminal region generating two fragments with potential neuroactivity (Fig. 3).

Interestingly, NPY has been shown to be neuroprotective and to promote the proliferation of neuronal precursor cells. NPY stimulates neuronal progenitor proliferation *in vitro* and in neurogenic brain regions [66, 90–92] (Fig. 2). Of these peptides that function as regulators of neurogenesis, NPY is widely expressed throughout the brain and has been shown to be an important participant in many neural functions, from social and feeding behavior [93] and circadian rhythm [94] to seizure control [95], learning, memory, and depression [96]. Interestingly, these latter functions are associated with activity of the hippocampus – where the DG shows particularly high NPY-like immunoreactivity [97] – and in this context, NPY's neurogenic effects may partially be responsible for the role of NPY in these widely varied functions [66]. NPY receptors are present in the hippocampus in areas of neurogenesis (Fig. 3).

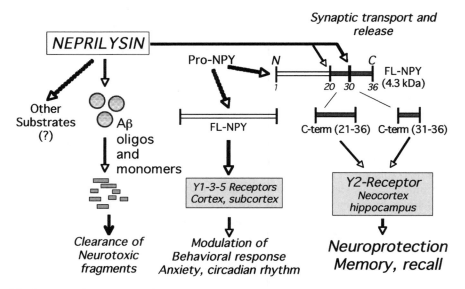

Fig. 3 Metabolic processing of neuropeptide Y (*NPY*) in the CNS and neuroprotection. Neprilysin (*NEP*) might ameliorate the neurodegenerative pathology in Alzheimer's disease (*AD*) by reducing Aβ and by generating protective CTF-NPY that may bind the Y2 receptor (*See Color Plates*)

Thus, NEP represents a unique example of a proteolytic enzyme with dual action: degradation (Aβ) and processing (NPY); both actions are neuroprotective. Therefore, understanding the effects of NEP on NPY and other neuropeptides might provide new information about the neuroprotective mechanisms of NEP in the mature CNS and in animal models of AD.

Acknowledgments This work was supported by NIH grants AG18440, AG022074, and AG10435. The funders had no role in study design, data collection and analysis, decision to publish, or preparation of the manuscript.

References

1. Ashford JW (2004) APOE genotype effects on Alzheimer's disease onset and epidemiology. J Mol Neurosci 23:157–165
2. Hebert LE, Scherr PA, Bienias JL et al. (2004) State-specific projections through 2025 of Alzheimer disease prevalence. Neurology 62:1645
3. Masliah E, Miller A, Terry R (1993) The synaptic organization of the neocortex in Alzheimer's disease. Med Hypotheses 41:334–340
4. Braak H, Braak E (1994) Morphological criteria for the recognition of Alzheimer's disease and the distribution pattern of cortical changes related to this disorder. Neurobiol Aging 15:355–356
5. Hof P, Morrison J (1991) Neocortical neuronal subpopulations labeled by a monoclonal antibody to calbindin exhibit differential vulnerability in Alzheimer's disease. Exp Neurol 111:293–301

6. Hof P, Morrison J (1994) The cellular basis of cortical disconnection in Alzheimer disease and related dementing conditions. In: Alzheimer Disease. Edited by Terry R, Katzman R, Bick K. New York: Raven Press; 197–230

7. Perry E (1995) Cholinergic signaling in Alzheimer disease: therapeutic strategies. Alzheimer Dis Assoc Disord 9(Suppl 2):1–2

8. Perry EK, Tomlinson BE, Blessed G et al. (1978) Correlation of cholinergic abnormalities with senile plaques and mental test scores in senile dementia. Br Med J 2:1457–1459

9. Trojanowski J, Schmidt M, Shin R-W et al. (1993) Altered *Tau* and neurofilament proteins in neurodegenerative diseases: diagnostic implications for Alzheimer's disease and Lewy body dementias. Brain Pathol 3:45–54

10. Masliah E, Terry R (1994) The role of synaptic pathology in the mechanisms of dementia in Alzheimer's disease. Clin Neurosci 1:192–198

11. Terry R, Masliah E, Salmon D et al. (1991) Physical basis of cognitive alterations in Alzheimer disease: synapse loss is the major correlate of cognitive impairment. Ann Neurol 30:572–580

12. Goldstein L, Ashford JW, Busciglio J et al. (2003) Live discussion. From here to there: AbetaPP as an axonal transport receptor – how could this explain neurodegeneration in AD. July 15, 2002. J Alzheimers Dis 5:483–489

13. Spires TL, Meyer-Luehmann M, Stern EA et al. (2005) Dendritic spine abnormalities in amyloid precursor protein transgenic mice demonstrated by gene transfer and intravital multiphoton microscopy. J Neurosci 25:7278–7287

14. Moolman DL, Vitolo OV, Vonsattel JP et al. (2004) Dendrite and dendritic spine alterations in Alzheimer models. J Neurocytol 33:377–387

15. Mucke L, Masliah E, Yu GQ et al. (2000) High-level neuronal expression of abeta 1–42 in wild-type human amyloid protein precursor transgenic mice: synaptotoxicity without plaque formation. J Neurosci 20:4050–4058

16. Sisodia S, Price D (1995) Role of the beta-amyloid protein in Alzheimer's disease. FASEB J 9:366–370

17. Selkoe D (1994) Cell biology of the amyloid β-protein precursor and the mechanisms of Alzheimer's disease. Ann Rev Cell Biol 10:373–403

18. Selkoe DJ (1994) Alzheimer's disease: a central role for amyloid. J Neuropathol Exp Neurol 53:438–447

19. Kamenetz F, Tomita T, Hsieh H et al. (2003) APP processing and synaptic function. Neuron 37:925–937

20. Sinha S, Anderson J, John V et al. (2000) Recent advances in the understanding of the processing of APP to beta amyloid peptide. Ann N Y Acad Sci 920:206–208

21. Selkoe DJ (1999) Translating cell biology into therapeutic advances in Alzheimer's disease. Nature 399(Suppl 6738):A23–A31

22. Sinha S, Anderson JP, Barbour R et al. (1999) Purification and cloning of amyloid precursor protein beta-secretase from human brain. Nature 402:537–540

23. Vassar R, Bennett BD, Babu-Khan S et al. (1999) Beta-secretase cleavage of Alzheimer's amyloid precursor protein by the transmembrane aspartic protease BACE. Science 286:735–741

24. Cai H, Wang Y, McCarthy D et al. (2001) BACE1 is the major beta-secretase for generation of Abeta peptides by neurons. Nat Neurosci 4:233–234

25. Luo Y, Bolon B, Kahn S et al. (2001) Mice deficient in BACE1, the Alzheimer's beta-secretase, have normal phenotype and abolished beta-amyloid generation. Nat Neurosci 4:231–232

26. DeKosky S, Scheff S (1990) Synapse loss in frontal cortex biopsies in Alzheimer's disease: correlation with cognitive severity. Ann Neurol 27:457–464

27. DeKosky ST, Scheff SW, Styren SD (1996) Structural correlates of cognition in dementia: quantification and assessment of synapse change. Neurodegeneration 5:417–421

28. Scheff SW, Price DA (2003) Synaptic pathology in Alzheimer's disease: a review of ultrastructural studies. Neurobiol Aging 24:1029–1046

29. Cotman C, Cummings B, Pike C (1993) Molecular cascades in adaptive versus pathological plasticity. In: Neurodegeneration. Edited by Gorio A. New York: Raven Press; 217–240

30. Masliah E, Mallory M, Alford M et al. (2001) Altered expression of synaptic proteins occurs early during progression of Alzheimer's disease. Neurology 56:127–129
31. Masliah E (2000) The role of synaptic proteins in Alzheimer's disease. Ann N Y Acad Sci 924:68–75
32. van Praag H, Schinder AF, Christie BR et al. (2002) Functional neurogenesis in the adult hippocampus. Nature 415:1030–1034
33. Gage FH, Kempermann G, Palmer TD et al. (1998) Multipotent progenitor cells in the adult dentate gyrus. J Neurobiol 36:249–266
34. Tatebayashi Y, Lee MH, Li L et al. (2003) The dentate gyrus neurogenesis: a therapeutic target for Alzheimer's disease. Acta Neuropathol (Berl) 105:225–232
35. Wen PH, Hof PR, Chen X et al. (2004) The presenilin-1 familial Alzheimer disease mutant P117L impairs neurogenesis in the hippocampus of adult mice. Exp Neurol 188:224–237
36. Chevallier NL, Soriano S, Kang DE et al. (2005) Perturbed neurogenesis in the adult hippocampus associated with presenilin-1 A246E mutation. Am J Pathol 167:151–159
37. Donovan MH, Yazdani U, Norris RD et al. (2006) Decreased adult hippocampal neurogenesis in the PDAPP mouse model of Alzheimer's disease. J Comp Neurol 495:70–83
38. Jin K, Galvan V, Xie L et al. (2004) Enhanced neurogenesis in Alzheimer's disease transgenic (PDGF-APPSw,Ind) mice. Proc Natl Acad Sci USA 101:13363–13367
39. Dong H, Goico B, Martin M et al. (2004) Modulation of hippocampal cell proliferation, memory, and amyloid plaque deposition in APPsw (Tg2576) mutant mice by isolation stress. Neuroscience 127:601–609
40. Rockenstein E, Mante M, Adame A et al. (2007) Effects of Cerebrolysintrade mark on neurogenesis in an APP transgenic model of Alzheimer's disease. Acta Neuropathol (Berl) 113:265–275
41. Feng R, Rampon C, Tang YP et al. (2001) Deficient neurogenesis in forebrain-specific presenilin-1 knockout mice is associated with reduced clearance of hippocampal memory traces. Neuron 32:911–926
42. Haughey NJ, Nath A, Chan SL et al. (2002) Disruption of neurogenesis by amyloid beta-peptide, and perturbed neural progenitor cell homeostasis, in models of Alzheimer's disease. J Neurochem 83:1509–1524
43. Wang R, Dineley KT, Sweatt JD et al. (2004) Presenilin 1 familial Alzheimer's disease mutation leads to defective associative learning and impaired adult neurogenesis. Neuroscience 126:305–312
44. Walsh DM, Selkoe DJ (2004) Oligomers on the brain: the emerging role of soluble protein aggregates in neurodegeneration. Protein Pept Lett 11:213–228
45. Glabe CC (2005) Amyloid accumulation and pathogensis of Alzheimer's disease: significance of monomeric, oligomeric and fibrillar Abeta. Subcell Biochem 38:167–177
46. Glabe CG, Kayed R (2006) Common structure and toxic function of amyloid oligomers implies a common mechanism of pathogenesis. Neurology 66:S74–S78
47. Lacor PN, Buniel MC, Chang L et al. (2004) Synaptic targeting by Alzheimer's-related amyloid beta oligomers. J Neurosci 24:10191–10200
48. Lesne S, Koh MT, Kotilinek L et al. (2006) A specific amyloid-beta protein assembly in the brain impairs memory. Nature 440:352–357
49. Luo JJ, Wallace W, Riccioni T et al. (1999) Death of PC12 cells and hippocampal neurons induced by adenoviral-mediated FAD human amyloid precursor protein gene expression. J Neurosci Res 55:629–642
50. Selkoe DJ (2000) The genetics and molecular pathology of Alzheimer's disease: roles of amyloid and the presenilins. Neurol Clin 18:903–922
51. Holtzman DM, Bales KR, Wu S et al. (1999) Expression of human apolipoprotein E reduces amyloid-ß deposition in a mouse model of Alzheimer's disease. J Clin Invest 103:R15–R21
52. Holtzman DM, Pitas RE, Kilbridge J et al. (1995) Low density lipoprotein receptor-related protein mediates apolipoprotein E-dependent neurite outgrowth in a central nervous system-derived neuronal cell line. Proc Natl Acad Sci USA 92:9480–9484
53. Nixon RA, Cataldo AM, Paskevich PA et al. (1992) The lysosomal system in neurons. Involvement at multiple stages of Alzheimer's disease pathogenesis. Ann N Y Acad Sci 674:65–88

54. Nixon RA, Wegiel J, Kumar A et al. (2005) Extensive involvement of autophagy in Alzheimer disease: an immuno-electron microscopy study. J Neuropathol Exp Neurol 64:113–122
55. Saito T, Iwata N, Tsubuki S et al. (2005) Somatostatin regulates brain amyloid beta peptide Abeta42 through modulation of proteolytic degradation. Nat Med 11:434–439
56. Iwata N, Tsubuki S, Takaki Y et al. (2001) Metabolic regulation of brain Abeta by neprilysin. Science 292:1550–1552
57. Iwata N, Tsubuki S, Takaki Y et al. (2000) Identification of the major Abeta1–42-degrading catabolic pathway in brain parenchyma: suppression leads to biochemical and pathological deposition. Nat Med 6:143–150
58. Selkoe DJ (2001) Clearing the brain's amyloid cobwebs. Neuron 32:177–180
59. Leissring MA, Farris W, Chang AY et al. (2003) Enhanced proteolysis of beta-amyloid in APP transgenic mice prevents plaque formation, secondary pathology, and premature death. Neuron 40:1087–1093
60. Eckman EA, Eckman CB (2005) Abeta-degrading enzymes: modulators of Alzheimer's disease pathogenesis and targets for therapeutic intervention. Biochem Soc Trans 33:1101–1105
61. Carson JA, Turner AJ (2002) Beta-amyloid catabolism: roles for neprilysin (NEP) and other metallopeptidases? J Neurochem 81:1–8
62. Spencer B, Marr R, Rockenstein EM et al. (2008) Long-term neprilysin gene transfer is associated with reduced levels of intracellular Abeta and behavioral improvement in APP transgenic mice. BMC Neurosci 9:109
63. Marr RA, Rockenstein E, Mukherjee A et al. (2003) Neprilysin gene transfer reduces human amyloid pathology in transgenic mice. J Neurosci 23:1992–1996
64. Iwata N, Mizukami H, Shirotani K et al. (2004) Presynaptic localization of neprilysin contributes to efficient clearance of amyloid-beta peptide in mouse brain. J Neurosci 24:991–998
65. Hong CS, Goins WF, Goss JR et al. (2006) Herpes simplex virus RNAi and neprilysin gene transfer vectors reduce accumulation of Alzheimer's disease-related amyloid-beta peptide in vivo. Gene Ther 13:1068–1079
66. Howell OW, Doyle K, Goodman JH et al. (2005) Neuropeptide Y stimulates neuronal precursor proliferation in the post-natal and adult dentate gyrus. J Neurochem 93:560–570
67. Skidgel RA, Erdos EG (2004) Angiotensin converting enzyme (ACE) and neprilysin hydrolyze neuropeptides: a brief history, the beginning and follow-ups to early studies. Peptides 25:521–525
68. Huang SM, Mouri A, Kokubo H et al. (2006) Neprilysin-sensitive synapse-associated amyloid-beta peptide oligomers impair neuronal plasticity and cognitive function. J Biol Chem 281:17941–17951
69. Akiyama H, Kondo H, Ikeda K et al. (2001) Immunohistochemical localization of neprilysin in the human cerebral cortex: inverse association with vulnerability to amyloid beta-protein (Abeta) deposition. Brain Res 902:277–281
70. Caccamo A, Oddo S, Sugarman MC et al. (2005) Age- and region-dependent alterations in Abeta-degrading enzymes: implications for Abeta-induced disorders. Neurobiol Aging 26:645–654
71. Reilly CE (2001) Neprilysin content is reduced in Alzheimer brain areas. J Neurol 248:159–160
72. Yasojima K, Akiyama H, McGeer EG et al. (2001) Reduced neprilysin in high plaque areas of Alzheimer brain: a possible relationship to deficient degradation of beta-amyloid peptide. Neurosci Lett 297:97–100
73. Yasojima K, McGeer EG, McGeer PL (2001) Relationship between beta amyloid peptide generating molecules and neprilysin in Alzheimer disease and normal brain. Brain Res 919:115–121
74. Oda M, Morino H, Maruyama H et al. (2002) Dinucleotide repeat polymorphisms in the neprilysin gene are not associated with sporadic Alzheimer's disease. Neurosci Lett 320:105–107
75. Sodeyama N, Mizusawa H, Yamada M et al. (2001) Lack of association of neprilysin polymorphism with Alzheimer's disease and Alzheimer's disease-type neuropathological changes. J Neurol Neurosurg Psychiatry 71:817–818
76. Clarimon J, Munoz FJ, Boada M et al. (2003) Possible increased risk for Alzheimer's disease associated with neprilysin gene. J Neural Transm 110:651–657

77. Dolev I, Michaelson DM (2004) A nontransgenic mouse model shows inducible amyloid-beta (Abeta) peptide deposition and elucidates the role of apolipoprotein E in the amyloid cascade. Proc Natl Acad Sci USA 101:13909–13914

78. Mohajeri MH, Wollmer MA, Nitsch RM (2002) Abeta 42-induced increase in neprilysin is associated with prevention of amyloid plaque formation in vivo. J Biol Chem 277:35460–35465

79. Shi J, Zhang S, Tang M et al. (2005) Mutation screening and association study of the neprilysin gene in sporadic Alzheimer's disease in Chinese persons. J Gerontol A Biol Sci Med Sci 60:301–306

80. Helisalmi S, Hiltunen M, Vepsalainen S et al. (2004) Polymorphisms in neprilysin gene affect the risk of Alzheimer's disease in Finnish patients. J Neurol Neurosurg Psychiatry 75:1746–1748

81. Sakai A, Ujike H, Nakata K et al. (2004) Association of the neprilysin gene with susceptibility to late-onset Alzheimer's disease. Dement Geriatr Cogn Disord 17:164–169

82. Lazarov O, Robinson J, Tang YP et al. (2005) Environmental enrichment reduces Abeta levels and amyloid deposition in transgenic mice. Cell 120:701–713

83. Rose JB, Rockenstein EM, Adame A et al. (2008) Role of NPY proteolysis in the mechanisms of neuroprotection mediated by neprilysin in a transgenic model of Alzheimer's disease. Submitted.

84. Marr RA, Guan H, Rockenstein E et al. (2004) Neprilysin regulates amyloid beta peptide levels. J Mol Neurosci 22:5–11

85. van Praag H, Christie BR, Sejnowski TJ et al. (1999) Running enhances neurogenesis, learning, and long-term potentiation in mice. Proc Natl Acad Sci USA 96:13427–13431

86. Ambree O, Leimer U, Herring A et al. (2006) Reduction of amyloid angiopathy and Abeta plaque burden after enriched housing in TgCRND8 mice: involvement of multiple pathways. Am J Pathol 169:544–552

87. Costa DA, Cracchiolo JR, Bachstetter AD et al. (2007) Enrichment improves cognition in AD mice by amyloid-related and unrelated mechanisms. Neurobiol Aging 28:831–844

88. Jankowsky JL, Xu G, Fromholt D et al. (2003) Environmental enrichment exacerbates amyloid plaque formation in a transgenic mouse model of Alzheimer disease. J Neuropathol Exp Neurol 62:1220–1227

89. Cracchiolo JR, Mori T, Nazian SJ et al. (2007) Enhanced cognitive activity – over and above social or physical activity – is required to protect Alzheimer's mice against cognitive impairment, reduce Abeta deposition, and increase synaptic immunoreactivity. Neurobiol Learn Mem 88:277–294

90. Hansel DE, Eipper BA, Ronnett GV (2001) Regulation of olfactory neurogenesis by amidated neuropeptides. J Neurosci Res 66:1–7

91. Hansel DE, Eipper BA, Ronnett GV (2001) Neuropeptide Y functions as a neuroproliferative factor. Nature 410:940–944

92. Hansel DE, May V, Eipper BA et al. (2001) Pituitary adenylyl cyclase-activating peptides and alpha-amidation in olfactory neurogenesis and neuronal survival in vitro. J Neurosci 21:4625–4636

93. Sokolowski MB (2003) NPY and the regulation of behavioral development. Neuron 39:6–8

94. Albers HE, Ferris CF (1984) Neuropeptide Y: role in light-dark cycle entrainment of hamster circadian rhythms. Neurosci Lett 50:163–168

95. Vezzani A, Sperk G, Colmers WF (1999) Neuropeptide Y: emerging evidence for a functional role in seizure modulation. Trends Neurosci 22:25–30

96. Redrobe JP, Dumont Y, St-Pierre JA et al. (1999) Multiple receptors for neuropeptide Y in the hippocampus: putative roles in seizures and cognition. Brain Res 848:153–166

97. Dumont Y, Martel JC, Fournier A et al. (1992) Neuropeptide Y and neuropeptide Y receptor subtypes in brain and peripheral tissues. Prog Neurobiol 38:125–167

Pore-Forming Neurotoxin-Like Mechanism for Aβ Oligomer-Induced Synaptic Failure

Luis G. Aguayo, Jorge Parodi, Fernando J. Sepúlveda, and Carlos Opazo

Abstract Cortical and hippocampal synapse densities are reduced in Alzheimer's disease (AD), and this strongly correlates with memory dysfunction. It is now believed that these changes in neuronal networking occur at the onset of AD and may lead to the neuronal loss displayed in later stages of the disease, which is characterized by severe cognitive and behavioral impairments. Mounting evidence indicates that amyloid-β (Aβ) oligomers are responsible for synaptic disconnections and neuronal death. One of the main consequences of Aβ oligomers interaction with neurons is an increase in intracellular Ca^{2+} concentration that could, when large enough, cause a marked alteration in ionic homeostasis. It has also been postulated that Ca^{2+} influx occurs when Aβ oligomers induce the opening of Ca^{2+} channels or the disruption of the plasma membrane. We recently found that the effects of Aβ oligomers on synaptic transmission are similar to pore-forming toxins, such as α-latrotoxin, a neurotoxin from the black widow spider. Here, we discuss evidence supporting a neurotoxin-like mechanism for the effects induced by Aβ oligomers on neuronal membranes, which could explain the alterations in the functionality of synapses in the central nervous system in AD that leads to major neurodegeneration with time of exposure to Aβ oligomers.

Abbreviations Aβ: amyloid-β peptide, AD: Alzheimer's disease; α-LTX: α-latrotoxin; AβPP: amyloid-β protein precursor; Ca^{2+}: calcium; LTP: long-term potentiation; pS: picoSiemen

1 Increase in Soluble Aβ Oligomers is a Key Factor for Alzheimer's Disease Onset

One of the main histopathological features of Alzheimer's disease (AD) is the presence of extracellular proteinaceous deposits in the brain, identified as senile plaques [1], which are enriched in amyloid-β (Aβ) peptide oligomers. It is widely

L.G. Aguayo (✉), J. Parodi, F.J. Sepúlveda, and C. Opazo
Department of Physiology, University of Concepción, Concepción Chile
e-mail: laguayo@udec.cl

R.B. Maccioni and G. Perry (eds.) *Current Hypotheses and Research Milestones in Alzheimer's Disease*. DOI: 10.1007/978-0-387-87995-6_2,
© Springer Science + Business Media, LLC 2009

accepted that AD onset can be initially triggered by interaction of Aβ oligomers with the brain parenchyma [1, 2]. However, the specific biochemical/structural characteristics of the Aβ oligomers that induce the neurotoxicity observed in AD have not been thoroughly identified, but it has been suggested that soluble Aβ oligomers (ranging from 17 to 56 kDa) [2, 3] are key determinants for neurotoxicity, including synaptic failure, observed in AD [2]. In agreement with this, the levels of soluble Aβ oligomers appear to correlate well with the severity of AD dysfunction [2, 4]. Interestingly, these soluble oligomers produced synaptic toxicity as expressed by inhibition of hippocampal long-term potentiation (LTP) *in vivo* and alterations in complex animal learning behaviors [2, 5]. These data strongly suggest that AD onset, probably associated to a mild synaptic dysfunction, occurs before amyloid plaque formation. However, the *mechanism (pre- or postsynaptic) by which Aβ oligomers cause synaptic dysfunction is largely unknown.*

Additionally, previous studies have shown that Aβ oligomers affect neuronal morphology and survival, and produce axonal and dendritic dystrophy [6]. These alterations seem to occur following amyloid deposition in the brain, indicating that accumulation of Aβ oligomers precedes the alterations in neuritic morphology [7]. Thus, these studies provide evidence suggesting a *decrease in neuronal networking in AD as a product of Aβ oligomers accumulation.* How this Aβ oligomers accumulation produces such a strong disruption in synaptic transmission in the central nervous system is currently under active investigation with the aim of discovering therapeutic targets and disease-modifying treatments.

2   Early Synaptic Alterations Precede AD Onset

It has been postulated that alterations in synaptic plasticity might be the primary failure responsible for the cognitive dysfunction in AD [8]. However, the scope and strength of studies supporting this challenging suggestion is only now being considered at the cellular and molecular level. It is currently known that synaptic transmission in the brain can be altered by specific and nonspecific mechanisms at pre- or postsynaptic sites.

In the case of Aβ, studies in hippocampal neurons treated with synthetic Aβ oligomers showed that it reduced the number of synaptic contacts and various pre- and postsynaptic proteins, thus suggesting extensive alterations in neuronal connectivity [9, 10]. In agreement, transgenic mice models overexpressing amyloid-β protein precursor (AβPP) showed a marked reduction in synaptophysin levels [11]. Interestingly, it was reported that synapse loss was highly correlated to neurological deficits observed in mild-to-severe stages of AD, supporting a direct link between cognitive functions and neurotransmission [12]. Furthermore, it was shown that early changes in synaptic morphology and markers such as synaptophysin correlate better to disease progression, suggesting that synaptic components are the most probable targets for the early neurotoxic actions of Aβ oligomers. Specifically, several proteins having well-defined functions in synaptic vesicle endocytosis, including AP2, AP180, dynamin, and synaptotagmin, have been reported to be extensively altered in AD [13].

How Aβ oligomers are able to produce this myriad of effects on synaptic proteins is unknown, but it is possible that these changes have a common triggering membrane mechanism. Nevertheless, these findings suggest that AD is associated with failure in the cellular machinery responsible for synaptic release and recycling. Additionally, alterations in synaptic proteins produced by the action of Aβ oligomers can explain its functional impact in models of cellular learning and memory, such as LTP [5]. *Here, we are proposing that Aβ oligomers affect synaptic transmission through its channel-forming properties (see below).* If Aβ oligomers cause synaptic transmission failure by pore formation, the following steps must occur for this mechanism to be demonstrated: (1) interaction of Aβ oligomers with neuronal membranes, (2) pore formation, (3) increase in intracellular calcium, (4) sustained increase in vesicular release, and (5) vesicular depletion (synaptic failure). We have found that Aβ oligomers produce several of these cellular events, as described below.

3   Calcium and Synaptic Dysfunction in AD

There is a current growing body of evidence suggesting the existence of a dysfunction in intracellular Ca^{2+} homeostasis in AD [14]. Prefibrillar Aβ oligomers have been shown to elevate Ca^{2+} in neurons. This increase in intracellular Ca^{2+} can follow receptor activation [15], modulation of voltage-activated Ca^{2+} channels [16], and influx via nonselective cations or by pore/channels formed by Aβ oligomers [17].

Analysis of the peptide secondary structure suggests the possibility of ion channel formation induced by membrane-bound Aβ oligomers [18]. The Aβ-pore/channel hypothesis was first proposed by Rojas and collaborators at the NIH using artificial membranes. They demonstrated the formation of pores with $Aβ_{1-40}$ that were highly cation-selective, allowing permeation of Ca^{2+}, Na^+, and Cs^+ [19]. These early studies in synthetic membranes were validated in membranes from hypothalamic cell lines [20]. Interestingly, cholesterol levels favored the formation of Aβ channels in artificial and hypothalamic membranes [20, 21]. Single channel measurements showed that the behavior of the $Aβ_{1-40}$-induced channels were exceptionally complex, in addition to their strong dependency on Cs^+ concentration and variability of single channel conductance (50–500 pS) [21]. Also, it was found that Zn^{2+}, known to bind Aβ in solution [22], blocked ion current flow [21], suggesting that the Aβ amyloid pore can be a pharmacological target. All together, the data suggest that Aβ oligomers do not form a unique, well-behaved type of ion channel, but they contribute to the formation of a complex multiple family of conducting pores [23]. Interestingly, using an "oligomer-enriched" form of Aβ, an increase was shown in lipid bilayer conductance, in the absence of unitary events, adding to the complex behavior of the peptide in the membrane [24].

In conclusion, it is evident that Aβ oligomers are able to increase the conductance in artificial membranes, but this has not been demonstrated in biologically relevant cell (neuron) membranes.

4   Proposed Neurotoxin-Like Mechanism for Aβ Oligomer-Induced Synaptic Failure

The cellular and molecular mechanisms that induce AD are largely unknown and deter development of effective disease-preventing/modifying therapies. The most accepted working hypothesis of AD is that excess of Aβ oligomers either (1) bind to membrane receptors affecting their functions [25], (2) interfere with signaling cascades [26], or (3) directly disrupt neuronal membranes causing pore formation thus leading to alterations in ionic homeostasis [21]. Although the latter is an attractive hypothesis because it could explain several effects of Aβ oligomers on brain synapses, it has not been documented to occur in brain neuronal membranes and this could be due to the high complexity of biological membranes, such as heterogeneity in native ion channels and receptors.

In an attempt to elucidate the mechanism by which Aβ oligomers induce synaptotoxicity, we undertook an experimental approach to characterize how Aβ oligomer affect synaptic transmission and compared these effects with those produced by neurotoxins known to form membrane pores. We found that the effects of Aβ oligomers, although at higher concentrations (nM vs pM), were very similar to those of pore-forming α-latrotoxin (α-LTX, 130 kDa) allowing us to suggest that its neurotoxicity was dependent on pore formation within the cell membrane. For example, similar to α-LTX [27, 28], we found that Aβ oligomers directly increased membrane conductance and intracellular calcium causing an early increase and a delayed failure in synaptic release (Fig. 1b).

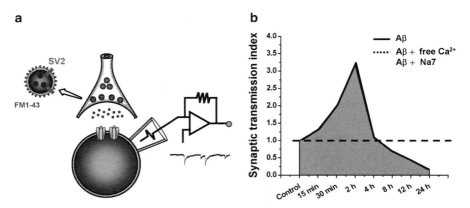

Fig. 1 Effects of $Aβ_{1-40}$ oligomers on synaptic activity of hippocampal neurons. **a** The scheme illustrates pre- and postsynaptic components of a central synapse. The vesicles are released in a calcium-dependent fashion. Presynaptic activity can be determined by the presence of vesicular proteins (SV2) or by the staining of synaptic vesicles with fluorescent probes such as FM1–43 (*red dots*). The postsynaptic membrane currents associated to the vesicular release and postsynaptic receptor density can be analyzed using the patch clamp technique. **b** Time-dependent biphasic effect of 500 nM Aβ oligomers on synaptic transmission in hippocampal neurons. The effects of Aβ oligomers were blocked by lowering extracellular calcium or by adding Na7 (*broken line*). The values were obtained from three independent experiments. (*See Color Plates*)

It has been recognized for several years that α-LTX can alter membrane permeability generating nonselective ionic pores [29]. Therefore, the mechanism for toxicity depends on attachment to the cell membrane and disruption of ionic permeability [27]. The majority of studies with α-LTX strongly support the idea that the main increase in intracellular calcium results from Ca^{2+} entry through nonselective cation channels formed by membrane-bound toxins [30]. Once bound to neuromuscular junction membranes, α-LTX forms oligomeric structures that stimulate exhaustive release of neurotransmitters [29]. One of the most distinguishing features of α-LTX on the synapse is that it produces a marked vesicular depletion [31]. Interestingly, we found that Aβ oligomers were able to mimic all of these effects in hippocampal neurons. For example, when examining the effects of nanomolar concentrations of $Aβ_{1-40}$ oligomers on the spontaneous synaptic activity in living hippocampal neurons, using patch clamp and fluorometric imaging (Fluo-3 and FM1–43) (Fig. 1a), we found that the effects of low concentrations of Aβ oligomers on synaptic transmission were biphasic, with a rapid facilitation followed by a delayed failure (Fig. 1b). We also found that the delayed synaptic failure correlated nicely with a decrease in several presynaptic proteins, such as SV2 (Fig. 2a and b). These results indicate that Aβ oligomers were able to produce a significant loss of connectivity in central neurons participating in learning and memory, which is in agreement with the idea that cognitive alterations in AD are associated to a synaptic failure [8, 12]. More importantly, blockade of the Aβ pore with a small peptide, previously shown to inhibit Aβ oligomers-induced increase on membrane permeability [32], protected the hippocampal neurons from synaptotoxicity, maintaining high levels of SV2 associated to neurotransmitter vesicles in the presence of Aβ oligomers (Figs. 1b and 2).

We propose that future studies of this membrane phenomenon will reveal that the target of Aβ oligomers are not another protein, *but will show that Aβ oligomers themselves are the cellular target thereby explaining the failure of pharmacological*

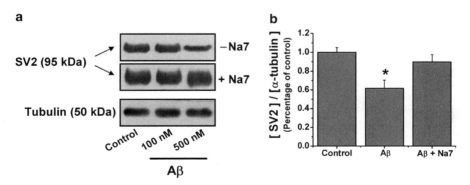

Fig. 2 The effect of $Aβ_{1-40}$ oligomers on vesicular SV2 level was blocked by Na7. **a** Western blots for SV2 obtained from hippocampal neurons incubated in the absence (control) or presence of $Aβ_{1-40}$ oligomers (100 and 500 nM) during 24 h. **b** Quantification of SV2 levels in the absence or presence of Na7 (200 nM). Note that Na7 blocked the reduction in SV2 induced by Aβ oligomers.

agents to modify the course of AD. Additionally, these studies should provide a new rationale for the development of drugs that block the Aβ pore and possibly interfere with AD onset.

Our data indicates that the effects of Aβ oligomers on intracellular Ca^{2+} play a key role in the alterations on synaptic transmission induced by the peptide, which is in agreement with previous studies involving this divalent cation on Aβ oligomers effects. For example, we found that the effects of Aβ oligomers on intracellular calcium and its associated synaptic transmission were largely attenuated by reducing the influx of calcium either by removal of this cation or by pharmacological means (Figs. 1b and 2). Thus, it is possible to conclude that the synaptic effects of Aβ oligomers are calcium-dependent [14] and able to be modulated.

Several questions concerning the mechanisms for Aβ oligomers insertion and perturbation of neuronal membranes should be resolved in future studies. For example, although formation of α-LTX pores seems to be mostly independent of membrane receptors, some membrane proteins could facilitate pore insertion [29]. Equivalent mechanisms may be true for the interaction of Aβ oligomers with neuronal membranes. Studies of the features of α-LTX using cryoelectron microscopy demonstrate that α-LTX penetrates the cell membrane and forms pores having a large diameter (10–25 Å) that facilitates the release of several neurotransmitters (e.g., norepinephrin, glutamate, and gamma amino butyric acid) by a nonvesicular efflux mechanism [29]. Such data is not available for Aβ-induced pores, but functional studies have indicated that Aβ pores are able to carry divalent cations [19, 33]. The inner diameters of Aβ pores, estimated by atomic force microscopy and molecular dynamics analysis [34, 35], have a similar range (15–20 Å). Therefore, they might permit the nonvesicular efflux of several metabolites, generating important changes in the metabolic cellular state. Because of the remarkable similarities in the mechanism of action between these pore-forming neurotoxins and Aβ oligomers (Table 1), we postulate that a pore-forming mechanism might explain how Aβ oligomers induce the synaptic dysfunction and neurodegeneration in AD (Fig. 3). We propose that Aβ oligomers might also bind to postsynaptic membranes causing their remodeling, but with a slower time course.

Table 1 Comparison between α-LTX- and Aβ oligomers-induced pores

	[1]α-LTX	[2]Aβ	References
MW of monomer	130 kDa	4 kDa	[[1], [1]29]
Proposed number of monomers/pore	4	≥12 and <24	[[1]29, [2]35]
Channel conductance (approximated)	Multiple levels, 100–300 pS	Multiple levels, 50–500 pS	[[2]21, [1]27]
Estimated inner pore diameter	10–25 Å	15–27 Å	[[1]29, [2]35]
Main cations transported	Ca^{2+}, Na^+	Ca^{2+}, Na^+, Cs^+	[[2]19, [1]27]
Effective protein concentration	<1–10 nM	100–500 nM	[[1]27, [2]33]
Onset of synaptic action	20 min	30 min	[[1]28, [2]33]
Acute enhancement of vesicular release	yes	yes	[[1]31, [2]33]
Delayed vesicular depletion	yes	yes	[[1]31, [2]33]

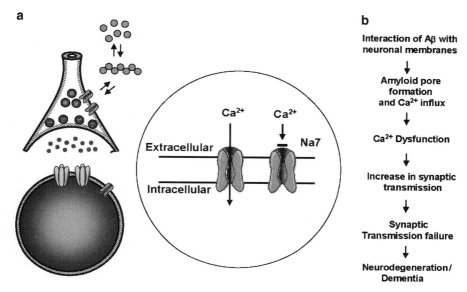

Fig. 3 Proposed hypothesis that explains the effect of Aβ oligomers on synaptic transmission. **a** Aβ oligomers bind to the membrane inducing the formation of pores in the pre- and postsynaptic membranes. The pores allow calcium to enter the cell and this event can be blocked by Na7. **b** Proposed series of events that leads to AD. Its initiation depends on oligomerization and membrane perturbations that lead to a calcium dysfunction and alterations in synaptic transmission. (*See Color Plates*)

5 Conclusions

We are currently studying the mechanisms that can explain the changes in intracellular calcium and synaptotoxicity following Aβ oligomers application to brain neurons by testing: (1) voltage-dependent calcium channels, (2) *N*-methyl D-aspartate (NMDA) receptors, and (3) membrane conductance following Aβ pore or channel formation. From this data, we expect to learn how Aβ oligomers inhibits synaptic transmission in brain neurons involved with learning and memory, with the aim of stopping or better still, reversing this process. The most interesting feature of Aβ oligomers on synaptic transmission is its biphasic action which leads to a strong synaptic failure that we interpret as a process of synaptic depletion. Interestingly, when the early effects of Aβ oligomers were blocked with low extracellular calcium, cadmium, or ruthenium red, the neurons did not show synaptic inhibition suggesting that the delayed inhibition was similar to the synaptic depletion induced by α-LTX.

According to previous and present evidence, three principal steps are involved in the neurotoxin-like mechanism for the action of Aβ oligomers to induce the early synaptic effects needed to trigger AD onset. First, Aβ oligomers have to bind to neuronal cell membranes, pre- and postsynaptic, long enough to ensure Aβ–Aβ interactions into the membrane. Second, Aβ forms oligomers into the cell membrane to allow the formation of a pore. Third, a sustained flow of cations (Ca^{2+}, Na^+) through the Aβ pores initiates the modification of synaptic activity, which leads to remodeling synaptic morphology. *We postulate that understanding the precise mechanism for each of these steps will greatly facilitate a pharmacological therapy urgently needed for the world-wide population affected by AD.*

Acknowledgments This work was supported by FONDECYT Grant No 1060368, Ring of Research PBCT ACT-04 (L.G.A. and C.O). We would like to thank Lauren Aguayo for her revision of the manuscript.

References

1. Masters CL, Simms G, Weinman NA et al. (1985) Amyloid plaque core protein in Alzheimer disease and Down syndrome. Proc Natl Acad Sci USA 82:4245–4249
2. Lesne S, Koh MT, Kotilinek L et al. (2006) A specific amyloid-β protein assembly in the brain impairs memory. Nature 440:352–357
3. Lambert MP, Barlow AK, Chromy BA et al. (1998) Diffusible, nonfibrillar ligands derived from Aβ1–42 are potent central nervous system neurotoxins. Proc Natl Acad Sci USA 95:6448–6453
4. McLean CA, Cherny RA, Fraser FW et al. (1999) Soluble pool of Aβ amyloid as a determinant of severity of neurodegeneration in Alzheimer's disease. Ann Neurol 46:860–866
5. Walsh DM, Townsend M, Podlisny MB et al. (2005) Certain inhibitors of synthetic amyloid β-peptide (Aβ) fibrillogenesis block oligomerization of natural Aβ and thereby rescue long-term potentiation. J Neurosci 25:2455–2462
6. Grace EA, Rabiner CA, Busciglio J (2002) Characterization of neuronal dystrophy induced by fibrillar Aβ: implications for Alzheimer's disease. Neuroscience 114:265–273
7. Meyer-Luehmann M, Spires-Jones TL, Prada C et al. (2008) Rapid appearance and local toxicity of Aβ plaques in a mouse model of Alzheimer's disease. Nature 451:720–724
8. Selkoe DJ (2002) Alzheimer's disease is a synaptic failure. Science 298:789–791
9. Kelly BL, Vassar R, Ferreira A (2005) Aβ-induced dynamin 1 depletion in hippocampal neurons. A potential mechanism for early cognitive decline in Alzheimer disease. J Biol Chem 280:31746–31753
10. Snyder EM, Nong Y, Almeida CG et al. (2005) Regulation of NMDA receptor trafficking by Aβ. Nat Neurosci 8:1051–1058
11. Hsia AY, Masliah E, McConlogue L et al. (1999) Plaque-independent disruption of neural circuits in Alzheimer's disease mouse models. Proc Natl Acad Sci USA 96:3228–3233
12. Terry RD, Masliah E, Salmon DP et al. (1991) Physical basis of cognitive alterations in Alzheimer's disease: synapse loss is the major correlate of cognitive impairment. Ann Neurol 30:572–580
13. Yao PJ (2004) Synaptic frailty and clathrin-mediated synaptic vesicle trafficking in Alzheimer's disease. Trends Neurosci 27:24–29
14. Mattson MP (2004) Pathways towards and away from Alzheimer's disease. Nature 430:631–639
15. Harkany T, Abraham I, Timmerman W et al. (2000) Aβ neurotoxicity is mediated by a glutamate-triggered excitotoxic cascade in rat nucleus basalis. Eur J Neurosci 12:2735–2745
16. Scragg JL, Fearon IM, Boyle JP et al. (2005) Alzheimer's amyloid peptides mediate hypoxic up-regulation of L-type Ca2+ channels. FASEB J 19:150–152
17. Kagan BL, Hirakura Y, Azimov R et al. (2002) The channel hypothesis of Alzheimer's disease: current status. Peptides 23:1311–1315

18. Durell SR, Guy HR, Arispe N et al. (1994) Theoretical models of the ion channel structure of Aβ-protein. Biophys J 67:2137–2145
19. Arispe N, Rojas E, Pollard HB (1993) Alzheimer disease Aβ-protein forms calcium channels in bilayer membranes: blockade by tromethamine and aluminum. Proc Natl Acad Sci USA 90:567–571
20. Kawahara M, Kuroda Y, Arispe N et al. (2000) Alzheimer's Aβ, human islet amylin, and prion protein fragment evoke intracellular free calcium elevations by a common mechanism in a hypothalamic GnRH neuronal cell line. J Biol Chem 275:14077–14083
21. Kawahara M, Arispe N, Kuroda Y et al. (1997) Alzheimer's disease Aβ-protein forms Zn2+-sensitive, cation-selective channels across excised membrane patches from hypothalamic neurons. Biophys J 73:67–75
22. Bush AI, Pettingell WH, Multhaup G et al. (1994) Rapid induction of Alzheimer Aβ amyloid formation by zinc. Science 265:1464–1467
23. Kourie JI, Henry CL, Farrelly P (2001) Diversity of Aβ protein fragment [1–40]-formed channels. Cell Mol Neurobiol 21:255–284
24. Kayed R, Sokolov Y, Edmonds B et al. (2004) Permeabilization of lipid bilayers is a common conformation-dependent activity of soluble amyloid oligomers in protein misfolding diseases. J Biol Chem 279:46363–46366
25. Bourin M, Ripoll N, Dailly E (2003) Nicotinic receptors and Alzheimer's disease. Curr Med Res Opin 19:169–177
26. Maccioni RB, Muñoz JP, Barbeito L (2001) The molecular bases of Alzheimer's disease and other neurodegenerative disorders. Arch Med Res 32:367–381
27. Van Renterghem C, Iborra C, Martin-Moutot N et al. (2000) α-latrotoxin forms calcium-permeable membrane pores via interactions with latrophilin or neurexin. Eur J Neurosci 12:3953–3962
28. Tsang CW, Elrick DB, Charlton MP (2000) α-Latroxin releases calcium in frog motor nerve terminals. J Neurosci 20:8685–8692
29. Orlova EV, Rahman M, Gowen B et al. (2000) Structure of α-latrotoxin oligomers reveals that divalent cation-dependent tetramers form membrane pores. Nat Struct Biol 7:48–53
30. Liu J, Misler S (1998) α-Latrotoxin alters spontaneous and depolarization-evoked quantal release from rat adrenal chromaffin cells: evidence for multiple modes of action. J Neurosci 18:6113–6125
31. Tzeng MC, Cohen RS, Siekevitz P (1978) Release of neurotransmitters and depletion of synaptic vesicles in cerebral cortex slices by α-latrotoxin from black widow spider venom. Proc Natl Acad Sci USA 75:4016–4020
32. Simakova O, Arispe NJ (2006) Early and late cytotoxic effects of external application of the Alzheimer's Aβ result from the initial formation and function of Aβ ion channels. Biochemistry 45:5907–5915
33. Parodi J, Sepúlveda FJ, Opazo C et al. (2008) Alzheimer amyloid-β causes a potent membrane perforation in brain neurons. New mechanism for the development and discovery of anti AD drugs. Manuscript in preparation
34. Lal R, Lin H, Quist A (2007) Aβ ion channel: 3D structure and relevance to amyloid channel paradigm. Biochim Biophys Acta 1768:1966–1975
35. Jang H, Zheng J, Nussinov R (2007) Models of Aβ ion channels in the membrane suggest that channel formation in the bilayer is a dynamic process. Biophys J 93:1938–1949

Interventions in Aging and Neurodegenerative Disease: Effects on Adult Stem Cells

Adam D. Bachstetter, Carmellina Gemma, and Paula C. Bickford

Abstract Throughout the entire life span, stem cells are present in many organs of our body and continue to produce new cells which are critical to maintain homeostasis and to repair damaged tissues. In the brain, stem cells generate new neurons through a process called neurogenesis. With age, stem cells lose their ability to generate new cells, although the number of stem cells remains constant over time. This may be due in part to cellular stresses such as inflammation, oxidative stress, and loss of trophic factors that accumulate with age. A better understanding of the regulatory factors which control neurogenesis is necessary in order to utilize the potential of the endogenous adult stem cells to treat the degenerative condition.

Abbreviations AD: Alzheimer's disease; CNS: central nervous system; GCL: granule cell layer; IL: interleukin; NSAID: nonsteroidal anti-inflammatory drug; NSCs: neural stem/progenitor cells; 6-OHDA, 6-hydroxydopamine; PD, Parkinson's disease; SGZ: subgranular zone; TNF: tumor necrosis factor-α

1 Introduction

Neurodegenerative diseases of aging such as Alzheimer's disease (AD) and Parkinson's disease (PD) are associated with a profound loss of synaptic plasticity and regionally selective cell loss. A similar loss of synaptic plasticity also occurs with normal aging, albeit not as severe and with less evidence for cell loss. The causes of the loss of synaptic plasticity are debatable but the results are clear to the patients who suffer from this devastating disease. As the population ages, there is a pressing need to develop therapeutic interventions for age-related neurodegenerative

A.D. Bachstetter, C. Gemma, and P.C. Bickford (✉)
Center of Excellence for Aging and Brain Repair, College of Medicine,
University of South Florida, Tampa, FL
e-mail: pbickfor@health.usf.edu

R.B. Maccioni and G. Perry (eds.) *Current Hypotheses and Research Milestones in Alzheimer's Disease*. DOI: 10.1007/978-0-387-87995-6_3,
© Springer Science + Business Media, LLC 2009

diseases, both for the patient and for those who will have the demanding responsibility to care for them. One aspect of neural plasticity is neurogenesis. It is now well accepted that neurogenesis occurs in at least two germinal centers in the brain. One of the neurogenic regions is the subventricular zone. The other known region found in the hippocampus is in the subgranular zone (SGZ) of the dentate gyrus, where neurogenesis has been found in humans as old as 72 years of age [1]. Addition of new neurons to the brain is complementary to synaptogenesis, which is another means of synaptic plasticity. Understanding the mechanisms that regulate neurogenesis is necessary to utilize this potential reservoir of synaptic plasticity to increase the quality of life of our aging population, with and without AD.

There are five phases of hippocampal neurogenesis:

1. The first phase is proliferation of the neural stem/progenitor cells (NSCs) which occurs in a region called the SGZ, which is roughly defined as a two-cell diameter band occurring on the hilus side of the granule cell layer (GCL).
2. The second phase is the survival of the proliferating NSCs. During this phase, the number of surviving neurons can vary greatly depending on the strain of animals used and can be as great as ~75% or as few as 25% of the amount of proliferating cells [2].
3. The third phase, occurring in concert with the second phase, is the differentiation of the newly born cells. In this phase, the majority of cells do become neurons, with a smaller percentage becoming astrocytes and oligodendrocytes.
4. The fourth phase involves migration of the neurons into the GCL, with most of the migration occurring around the first week [3].
5. Finally, the fifth phase involves the functional maturation of the neurons in the GCL which occurs around 4 weeks of age, but some cells may take weeks or even months longer to fully mature [4].

It has also been demonstrated that those adult born neurons that survive after 4 weeks will likely be present at least 11 months later [3]. The majority of the decrease in neurogenesis with age appears to occur mostly in the first phase where there is a decrease in proliferation with advanced age [5]. Survival of the newly born cells appears to be unaffected by age, while maturation of the cells particularly in developing a mature neuronal phenotype and migrating into the GCL does seem to be affected by age [5].

Beyond the limited effect of synaptic plasticity in the two germinal centers in the brain, neurogenesis from endogenous NSC may provide an alternative to transplantation of stem cells as a means to replace damaged neural tissue after brain injury, such as stroke, or as a result of a neurodegenerative condition. A growing body of research shows that neurogenesis may occur in "nonneurogenic" regions as a result of a stroke or neurodegenerative disease [6]. There is also the hope of recruiting NSCs from the neurogenic regions of the brain to replace damaged cells after injury [7]. Before either strategy of using neurogenesis as a therapeutic source of new cells can be implemented, a better understanding of the regulation of neurogenesis is necessary.

In many pathological conditions, including AD and PD, neurogenesis is dramatically affected by the pathology [8–11]. Neurogenesis shows a significant decrease with age [12–14]. The majority of the suppression of neurogenesis with age appears not to be NSCs autonomous but is more a function of the microenvironment, as any number of environmental alterations can increase or decrease neurogenesis (reviewed in [15]). Furthermore, the pool of NSC appears to be intact with respect to the total numbers of available cells [16], providing more evidence that that the neurogenic niche is at least partly responsible for the decrease in neurogenesis with age.

That cellular senescence occurs with age has been known since the 1960s [17], but the importance of the cellular senescence within the aged niche has only recently become an area of active interest. A clear example of the importance of the extrinsic or systemic influence on the stem cell niche was demonstrated in the stem cells that are found in the muscle, called satellite cells. Like the NSCs, the satellite cells in the muscle lose the potential to regenerate damaged tissue with age. In an elegant experiment, when aged rats were exposed to the systemic environment of a young rat by parabiosis, the satellite cells were rejuvenated in the aged rats as demonstrated by an increase in the proliferation rate. Conversely, in young rats, the exposure to the circulation of the aged rats caused a decrease in the regenerative potential of the satellite cells [18], again supportive of an extrinsic/circulating factor that is influencing the proliferation of the stem cells in the aged animals. It is not clear whether the mechanism involved in the effect in the muscle would hold true in the brain, but the implication is that the aged environment is detrimental to stem cell function. This also holds true for even the most pluripotent of stem cells: the embryonic stem cells. When embryonic stem cells are transplanted into aged tissue they are not able to repair damaged tissue as well as when transplanted into young tissue [19]. These aforementioned examples demonstrate the importance of understating the biology of the aged stem cell niche and may provide insight into why some of the clinical trials which used cell transplantation for neurodegenerative disease have not been as effective as was hoped.

Thus, it appears that for a stem cell-based therapy to become efficacious, a better understanding of the aged niche and how the niche regulates the stem cell potential will be necessary. A multitude of changes that occur to the microenvironment of the aged niche may be related to the decrease in neurogenesis. One potential change is a decrease in a number of trophic factors including BDNF, VEGF, IGF-1, and FGF-2 [20, 21]. There is also an increase in corticosteroid levels with age [22]. A hallmark of aging as well as AD and PD is an increase in inflammation. While a loss of trophic factors and increased corticosteroids are important contributes to the aged niche and have been extensively reviewed [5], the focus of this chapter is limited to the role of inflammation in regulating stem cell function in the aged niche.

2 Neuroinflammation in Aging and Disease

Inflammation is an active process with the purpose of removing or inactivating potentially damaging agents. Following removal of the "danger signal," a second pathway is initiated with the role of tissue remodeling. In the central nervous system (CNS), the inflammatory process must be well controlled. Since a majority of the CNS lacks the potential to replace lost cells, an inflammatory insult could be devastating, resulting in neural tissue loss. In general, inflammation is a good thing with the primary result of removing the noxious agent and remodeling the adjacent tissue. When inflammation is not well regulated following response to "danger signals," a chronic pathology will result.

It is known that with aging and age-related neurodegenerative diseases, namely, AD and PD, there is a state of chronic inflammation. The cause of the chronic inflammation in aging is not clear. It was initially believed that in AD, inflammation was exacerbating the disease. A surprising experiment inducing inflammation in a mouse model of amyloid pathology through the use of lipopolysaccharide resulted in a paradoxical reduction in amyloid-β [23]. These results seem to contradict the common hypothesis that chronic inflammation is part of the disease progression. This contradiction comes from a shortcoming in the inflammation hypothesis, namely, inflammation is a negative that should be eliminated. The immune system is not a mistake of evolution. The majority of time the immune system protects us from our environment. In the CNS, the immune response has to be well controlled for the response to be beneficial. Two recent examples of the beneficial role of the immune system were demonstrated in an animal model of PD.

In the first experiment (Fig. 1a and b), rats were fed a diet enriched with blueberry or *Spirulina* prior to a 6-hydroxydopamine (6-OHDA) lesion. At 1 week after the lesion, all the animals irrespective of diet had a similar lesion volume. At 4 weeks after the lesion, the animals that were fed the diet enriched with blueberry or *Spirulina* had a significant reduction in the size of the lesion. The surprising result was the improved recovery at 4 weeks after the lesion may have been due to a more robust immune response at 1 week after the lesion. As shown in Fig. 1b, rats on the blueberry or *Spirulina* diets had a significant increase in the number of activated microglia at 1 week after the lesion and this was followed by a significant decrease in microglia activation at 4 weeks after the lesion [24]. These results suggest that a robust early microglia response is beneficial in removing the dead and dying cells and remodeling the remaining tissue.

In a follow-up experiment (Fig. 1c and d), the immediate immune response versus the chronic response was clearly elucidated. Tumor necrosis factor-α (TNFα) is a key proinflammatory cytokine that is produced by a variety of cell types but in PD patients is particularly associated with activated microglia [25]. Using the same rat model of PD as the previous experiment, an antisense TNF-specific oligodeoxyribonucleotide was given either immediately after the insult or beginning 5 days after the insult. Blockade of immediate TNFα effects was detrimental, but blocking the production of TNFα at the later time point was beneficial (Fig. 1c) [26]. As was seen with the diet supplementation study, the early activation of the immune response appears to be necessary and beneficial, whereas a state of chronic inflammation is detrimental.

One way in which chronic inflammation is detrimental to tissue remodeling after injury is by decreasing the regenerative potential of stem cells. Two seminal studies a number of years ago showed that inflammation tightly regulates

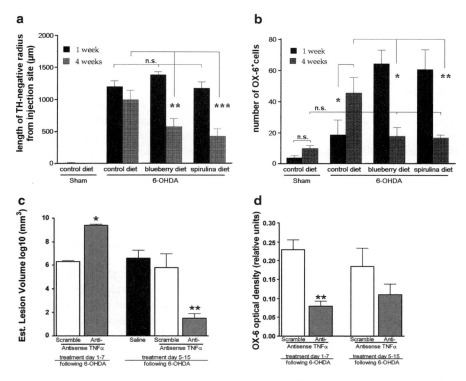

Fig. 1 Neuroinflammation is protective after toxic insult if the response is well controlled. **a** Following a 6-hydroxydopamine (*6-OHDA*) lesion in the striatum, a loss of tyrosine hydroxylase (*TH*) immunoreactivity was seen in all lesioned animals at 1 week. At 4 weeks after the lesion, a significant reduction in lesion volume was found in those rats that were treated with the blueberry- or *Spirulina*-enriched diets. **b** The protective effect at 4 weeks in rats that were fed a blueberry- or *Spirulina*-enriched diet may have been due to a more robust early microglia response at 1 week, as determined by the number of OX-6-positive microglia. The early microglia response at 1 week was followed by a return to nonlesions numbers of OX-6-positive microglia in the enriched diet groups at 4 weeks. **c** When an early immune response was blocked by a tumor necrosis factor-α (*TNFα*) antisense for the first 7 days after the lesion, a larger lesion volume was found compare to the control. However, inhibiting TNFα for 10 days starting 5 days after the lesion resulted in a beneficial effect. **d** A reduction in OX-6-positive microglia was found in the TNFα antisense-treated animals. The decrease in microglia by TNFα antisense during the first 7 days following the lesion may have blocked the early microglia response which was found to be protective in the animals fed a blueberry- or *Spirulina*-enriched diet. However, blocking TNFα during days 5–15 appeared to stop the chronic inflammatory response which was likely detrimental in the control rats that did not receive the TNFα antisense or the enriched diets. $^*P < 0.05$, $^{**}P < 0.01$, and $^{***}P < 0.001$

neurogenesis [27, 28]. Since these studies, a number of laboratories have been interested in how inflammation regulates neurogenesis. The results are fairly clear and reproducible, namely neuroinflammation will result in diminished neurogenesis, and decreasing the neuroinflammation will restore neurogenesis. What is not clear is how inflammation regulates neurogenesis.

3   The Role of Cytokines in Regulating Neurogenesis

Interplay between the immune system and the CNS underlies many of the neuro-physiological changes that occur with aging. Cytokines are a class of polypeptides expressed at low levels in healthy tissue that are rapidly induced in response to trauma or immune challenge. Levels of specific cytokines expressed in the brain increase as a function of age, even in the absence of a pathological stimulus. For example, there is a progressive increase in the expression of interleukin (IL)-1 and microglia activation with aging in neurologically intact patients [29, 30]. IL-6 levels also increase in the mouse brain with advancing age. In the cerebellum of aged rats, tumor necrosis factor-α (TNFα) gene expression is dramatically increased compared to young rats, and this increase is prevented by a diet rich in antioxidants [31]. Immune response-related molecules and their receptors are expressed throughout the brain, and recent research suggests that brain-derived immune factors disrupt normal physiology and contribute to cognitive and behavioral dysfunction in neurological disease [32–35].

IL-1β is one of the main inflammatory cytokines found in the CNS. IL-1β is constitutively expressed in the brain, synthesized by neuronal and/or glial cells, and released in response to a variety of stimuli, including immune system activation[36, 37]. IL-1β is a proinflammatory cytokine initially synthesized as an inactive precursor that is cleaved by caspase-1 to generate the biologically mature 17-kDa form. IL-1β affects virtually every cell type by binding to a high-affinity receptor, IL-1RI [37, 38]. The IL-1β receptor expression is high in the hippocampus as indicated by binding studies [39, 40]. The IL-1 family comprises three known ligands: IL-1α, IL-1β, and IL-1ra. The biological activity of IL-1β is dependent on its interaction with IL-1RI and recruitment of the IL-1 receptor accessory protein (IL-1Racp) [41]. IL-1ra binds to IL-1RI but fails to associate with IL-1Racp, thereby acting as a highly selective competitive receptor inhibitor. The only known function of IL-1ra is to prevent the biological activity of IL-1.

The potential role of IL-1β in the mediation of an age-related reduction in neurogenesis and learning and memory has been examined. In one study, the nonsteroidal anti-inflammatory drug (NSAID) sulindac was administered to aged rats. In this study, we found that sulindac treatment resulted in a decline in IL-1β levels and a reversal in the age-related deficits in radial arm water maze performance and contextual fear conditioning [42] (Fig. 2a). In a second experiment to determine if IL-1β is involved in the reversal of age-related cognitive deficits, we used an enzymatic inhibitor of caspase-1. Caspase-1 cleaves immature IL-1β to produce the mature, active form. The caspase-1 inhibitor Ac-YVAD-CMK was given for 28 days via an osmotic minipump into the left lateral ventricle to 4-month-old and 20-month-old male Fischer-344 rats. After 20 days of caspase-1 inhibition, the rats were trained in the contextual fear conditioning task. On day 22, they were tested for the hippocampus-dependent form of memory. Using the caspase-1 inhibitor, we found a similar improvement in the memory test as was seen with the NSAID (Fig. 2b) [43]. In a follow-up experiment, we hypothesized that the caspase-1 inhibition may be improving hippocampal-dependent memory by reversing the age-related decline

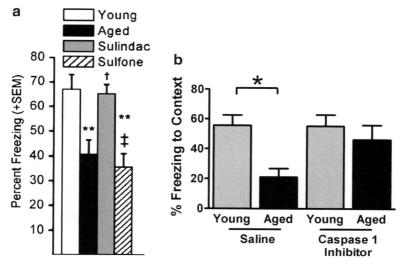

Fig. 2 Age-related chronic neuroinflammation impairs cognitive function. Decreased neuroinflammation in the aged brain by a nonsteroidal anti-inflammatory drug (*NSAID*) (**a**) or caspase-1 inhibitor (**b**) restored cognitive function in aged rats. **a** Sulindac, an NSAID that is a nonselective cyclooxygenase (*COX*) inhibitor, was shown to reverse the age-related deficits in contextual fear conation in 20-month-old male fisher rats, while sulfone a non-COX active metabolite was not able to. The improvement in memory was correlated with the reduction in interleukin (*IL*)-1b. There was no age-related effect on freezing to the novel context; $^{**}P < 0.005$ from young, $^{\dagger}P < 0.005$ from aged/control, and $^{\ddagger}P < 0.001$ from aged/sulindac. **b** Using a caspase-1 inhibitor which blocks the formation of active IL-1β resulted in a similar reversal of age-related deficits in contextual fear as the NSAID treatment. $^{*}P < 0.05$

in neurogenesis. Following the same experimental methods as the previous study, we found that caspase-1 inhibition did result in an increase in neurogenesis in the aged rats (Fig. 3) [44].

As expected caspase-1 inhibition dramatically reduced IL-1β levels, but the effect was not limited to IL-1β. In addition, caspase-1 inhibition also decreased TNFα and microglia activation. Furthermore, there was also increase in the anti-inflammatory cytokines IL-10, as well as an increase in IGF-1 [43, 44]. So while the target for caspase-1 inhibition was to decrease IL-1β levels, many inflammatory pathways were altered by caspase-1 inhibition. This is in part due to the interdependency of the cytokines and their cascading effect. Nevertheless, it has been shown that the NSCs do express the receptors for both IL-1β and TNFα, thereby having the potential to effect proliferation, development, or survival of the NSC [45–47].

IL-1β is a potent suppressor of neurogenesis but it is not completely clear how IL-1β achieves this effect. IL-1β is able to induce apoptosis in NPC phosphorylation of the SAPK/JNK pathway [45], thereby having a direct effect on the survival of the newborn neurons. The NF-κB/IκK pathway also seems to be critical in regulating the effects of IL-1β on proliferation, possibly by decreasing cyclin D1 expression [47]. IL-1β also seemed to slightly favor differentiation into an astrocyte linage as determined by the marker GFAP [45].

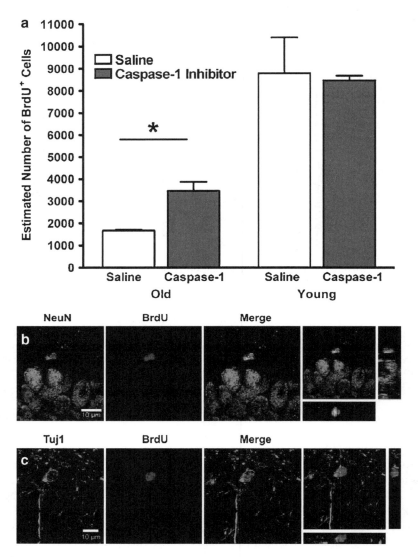

Fig. 3 Age-related neuroinflammation causes a decrease in hippocampal neurogenesis. **a** Aging results in a significant decrease in neurogenesis, as was seen by the decrease in BrdU-postive cells in the subgranular zone/granule cell layer (*SGZ/GCL*) of aged fisher 344 rats compare to young adult rats. The age-related decrease in neurogenesis was attenuated by 28 days of intracerebroventricular infusion of Ac-YVAD-CMK, a capase-1 inhibitor (*$P < 0.05$). Caspase-1 inhibition blocks the production of pro-IL-1β to the active form of interleukin (*IL*)-1β. The results demonstrate that at least part of the decrease in hippocampus neurogenesis with age is due to elevated cytokine levels. Representative confocal micrograph of (**b**) BrdU-labeled cells (red) and NeuN+ (green) and (**c**) BrdU+ (red) and TUJ1+ (green), confirms that the increase in BrdU+ cells did represent an increase in neurogenesis (*See Color Plates*)

IL-1β is not alone in being able to directly affect NSC. Recently, Iosif (2006) has shown that the inflammatory cytokine TNFα can also assert a direct effect on NSC proliferation. TNF receptor 1 (TNF-R1) appears to have a regulatory function by blocking proliferation during inflammation. This action occurs directly at the NSCs which express both TNF-R1 and TNF-R2. TNF-R2 appears to play a neuro-protective role, although its function is a little less clear than that of TNF-R1 [46].

4   Effect on Microglia

The main cell type in the CNS that is responsible for immunity is the microglia. While astrocytes and neurons can make inflammatory mediators, the microglia are the main source of inflammatory cytokines.

Microglia are always surveying the microenvironment, and once they sense the appropriate queues such as neuronal damage, the cells will hone to the site of damage. This initial phase of activation appears to be very beneficial in protecting the brain, but if the activation continues unabated then more damage can result [24]. It appears that this effect is due to the delicate nature of the CNS and neurons' inability to regenerate in most regions of the brain. As the resident immune cells in the CNS, microglia constitutively express surface receptors that trigger or amplify the innate immune response, including Toll-like receptors, complement receptors, cytokine receptors, chemokine receptors, major histocompatibility complex II, and others [48]. However, the role of microglia is not destructive. Upon detection of homeo-static disturbance, microglia rapidly respond by inducing a protective immune response. The protective immune response begins with a transient upregulation of inflammatory molecules, including proinflammatory cytokines such as TNF, IL-1, and IL-6, and IL-12 [49, 50]. This is followed by a protective phase that is immu-nomodulatory and neuroprotective. The protective phase includes neurotrophic factors such as BDNF, GDNF, and IGF-1 [51–53]. Thus, microglia remove cells damaged from acute injury and protect CNS functions.

Microglia are neither proneurogenic nor antineurogenic but their influence on neurogenesis is dependent on their activation state of either "classically" activated or "alternatively" activated [54, 55] (Fig. 4). Much like what has been proposed for peripheral macrophages, microglia are very pleiotropic. Microglia can become "classically" activated as defined by the release of proinflammatory cytokines (e.g., TNFα and IL-1β). Once in this "classically" activated, pro-inflammatory state, microglia are associated with further production of these pro-inflammatory cytokines, reactive oxygen species, chemokines, and matrix metalloproteases, resulting in cell death of invading cells and tissue destruction, and type-I inflamma-tion. A second type of microglia are those that are activated by such things as IL-4 and TGFβ, and are called "alternatively" activated or in an "M2" state (following the TH1/TH2 classification of T-helper cells). Compare to the "M1" or "classically" activated state, microglia in the "M2" state can be protective [50]. When macro-phages/microglia are in this "M2" state, there is little release of proinflammatory

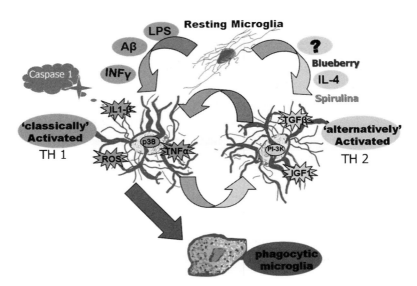

Fig. 4 Microglia: protective or harmful? Microglia are normally in a resting state in which they are actively surveying the microenvironment of the brain. The microglia are resting in the sense that they are not performing effector functions such as producing inflammatory mediators like interleukin (*IL*)-1β and tumor necrosis factor-α (*TNFα*). When microglia are producing inflammatory mediators, the microglia would be considered in a "classically activated state" or "TH1" state. Microglia can also become "alternatively activated" in such a way that they produce growth factors, such as IGF-1 and TGFβ. The "alternatively activated" microglia can support tissue remodeling and repair. Beyond releasing signaling molecules, microglia also have an important role in phagocytosis. The role of microglia, as protective or harmful, depends upon the ability of the microglia to switch from the different activation states at the appropriate time. Understanding how and when to turn microglia "on" or "off" is an important future direction of research. This is especially the case with aging where microglia are most needed to remodel and repair and to remove damaged cells and misfolded proteins. With age, microglia may lose the ability to perform these important effector functions making the aged brain more susceptible to injury and insult (*See Color Plates*)

cytokines and they are resistant to activation by agents such as LPS. In this alternative state, macrophages promote extracellular matrix formation and angiogenesis. A third type of microglia is one that becomes dysfunctional with age. The senescent microglia are not able to respond appropriately to stimuli and may exacerbate neuroinflammation [56, 57]. When microglia are in what is believed to be a "senescent"/ "classically" activated state, their numbers do seem to regulate neurogenesis (Fig. 5c) [58], although how this occurs is unclear.

5   The Role of Adaptive Immunity in Regulating Neurogenesis

It has been clearly shown that in the CNS there is a marked increase in inflammatory activity associated with aging; it is less clear what causes this inflammatory state [59–61]. It appears that there may be a correlation between inflammation in

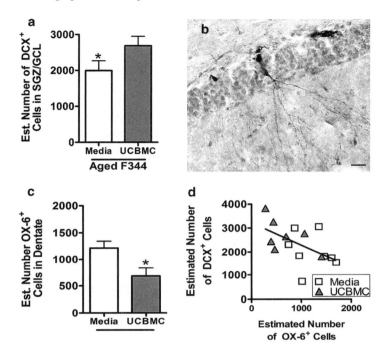

Fig. 5 Umbilical cord blood promotes neurogenesis. **a** 10^6 Umbilical cord blood cells were injected intravenously into 20-month-old F344 rats. In the aged rats, a significant increase in neurogenesis was found ($^*P < 0.05$), as shown by the increase in the number of doublecortin+ cells in the subgranular zone/granule cell layer (*SGZ/GCL*). **b** Representative photomicrograph of staining. **c** The intravenous injection of umbilical cord blood cells caused a significant decrease in the number of activated microglia ($^*P < 0.05$). **d** A negative correlation was found such that a decrease in the number of activated microglia was correlated with more neurogenesis in the aged rats

the CNS and a decrease in immunological function in the periphery [62, 63]. This correlation, while interesting, adds little to the explanation of what is occurring in the CNS if we maintain the view that the CNS is immunologically privileged, where overzealous immune cells are blocked from causing havoc. It appears that the "immune privilege" of the CNS is not completely true. As it turns out the CNS does have a reservoir to drain antigen in the cervical lymph nodes [64]. The recruitment of leukocytes can also take place in postcapillary venules [65]. Normally, in a young adult in the absence of disease, there is a clear demarcation between the peripheral immune system, innate and adaptive, and the CNS immune system with its "lone sentinel," the microglia. As can be seen following an acute stimulation such as a traumatic injury or ischemia, the protective status of the CNS is revoked and peripheral immune cells (dendritic cells and T cells) can enter into the CNS. The consequences of the invading peripheral immune cells is to either clear debris and rebuild damaged tissue or exacerbate the injury. The result is dependant on a well-controlled response that appears to be governed by T cells [66].

Injury is not the only time that this immune privilege is revoked; in aging, you can also find invasion of T cells and dendritic cells [67]. Occurring as early as 12 months of age in rats, dendritic cells and T cells can be found in the brain. These peripheral immune cells, which are absent in young adult rats, are found widely distributed throughout the aged rat brain, and are particularly associated with white matter tracks [67, 68] What causes this invasion of peripheral immune cells is not entirely known. It has been shown that there are changes to structure and function of the blood–brain barrier as a result of age, which causes an increase in permeability [69–71]. It has also been demonstrated that the effects of a T-cell response may be through altering the activity of the microglia and this response is important for maintain neurogenesis [55, 56, 72, 73].

Recently, we have demonstrated that in intravenous injection of human umbilical cord blood was able to restore the potential of the aged NSCs (Fig. 5) [58]. The effect of the umbilical cord blood cells was believed to act by an interaction in the peripheral circulation. Interestingly, coinciding with the increase in neurogenesis, a decrease in microglia activation was also found. Furthermore, there was a negative correlation found whereby those animals that had decreased numbers of activated microglia had an increase in neurogenesis. The results do suggest that circulating immune cells can have an effect on the CNS.

6   Conclusion

Neurogenesis is an important means of synaptic plasticity that is diminished with increasing age. During aging and exacerbated in AD, a state of chronic inflammation occurs. Inflammation, while not the only cause for the decrease in neurogenesis with age, is very important in regulating neurogenesis. To harness the potential of neurogenesis to help in brain repair, a better understanding of the mechanism of how inflammation is regulating neurogenesis is necessary. Furthermore, while high levels of cytokines do inhibit neurogenesis, it should be remembered that inflammation is beneficial when it is well controlled. Therapeutic interventions that help to regain the immune balance that is lost with age should be an important consideration in developing treatments for aging and age-related neurodegenerative diseases.

References

1. Eriksson PS, Perfilieva E, Bjork-Eriksson T et al. (1998) Neurogenesis in the adult human hippocampus. Nat Med 4: 1313–1317
2. Kempermann G, Kuhn HG, Gage FH (1997) Genetic influence on neurogenesis in the dentate gyrus of adult mice. Proc Natl Acad Sci USA 94: 10409–10414
3. Kempermann G, Gast D, Kronenberg G, Yamaguchi M, Gage FH (2003) Early determination and long-term persistence of adult-generated new neurons in the hippocampus of mice. Development 130: 391–399

4. van Praag H, Schinder AF, Christie BR, Toni N, Palmer TD, Gage FH (2002) Functional neurogenesis in the adult hippocampus. Nature 415: 1030–1034
5. Drapeau E, Nora Abrous D (2008) Role of neurogenesis in age-related memory disorders. Aging Cell 7(4): 569–589
6. Kokaia Z, Lindvall O (2003) Neurogenesis after ischaemic brain insults. Curr Opin Neurobiol 13: 127–132
7. Bernal GM, Peterson DA (2004) Neural stem cells as therapeutic agents for age-related brain repair. Aging Cell 3: 345–351
8. Zhang C, McNeil E, Dressler L, Siman R (2007) Long-lasting impairment in hippocampal neurogenesis associated with amyloid deposition in a knock-in mouse model of familial Alzheimer's disease. Exp Neurol 204: 77–87
9. Verret L, Jankowsky JL, Xu GM, Borchelt DR, Rampon C (2007) Alzheimer's-type amyloidosis in transgenic mice impairs survival of newborn neurons derived from adult hippocampal neurogenesis. J Neurosci 27: 6771–6780
10. Nuber S, Petrasch-Parwez E, Winner B et al. (2008) Neurodegeneration and motor dysfunction in a conditional model of Parkinson's disease. J Neurosci 28: 2471–2484
11. Hoglinger GU, Rizk P, Muriel MP et al. (2004) Dopamine depletion impairs precursor cell proliferation in Parkinson disease. Nat Neurosci 7: 726–735
12. Kempermann G, Gast D, Gage FH (2002) Neuroplasticity in old age: sustained five-fold induction of hippocampal neurogenesis by long-term environmental enrichment. Ann Neurol 52: 135–143
13. Kronenberg G, Bick-Sander A, Bunk E, Wolf C, Ehninger D, Kempermann G (2006) Physical exercise prevents age-related decline in precursor cell activity in the mouse dentate gyrus. Neurobiol Aging 27: 1505–1513
14. Kuhn HG, Dickinson-Anson H, Gage FH (1996) Neurogenesis in the dentate gyrus of the adult rat: age-related decrease of neuronal progenitor proliferation. J Neurosci 16: 2027–2033
15. Zhao C, Deng W, Gage FH (2008) Mechanisms and functional implications of adult neurogenesis. Cell 132: 645–660
16. Hattiangady B, Shetty AK (2008) Aging does not alter the number or phenotype of putative stem/progenitor cells in the neurogenic region of the hippocampus. Neurobiol Aging 29: 129–147
17. Hayflick L, Moorhead PS (1961) The serial cultivation of human diploid cell strains. Exp Cell Res 25: 585–621
18. Conboy IM, Conboy MJ, Wagers AJ, Girma ER, Weissman IL, Rando TA (2005) Rejuvenation of aged progenitor cells by exposure to a young systemic environment. Nature 433: 760–764
19. Carlson ME, Conboy IM (2007) Loss of stem cell regenerative capacity within aged niches. Aging Cell 6: 371–382
20. Hattiangady B, Rao MS, Shetty GA, Shetty AK (2005) Brain-derived neurotrophic factor, phosphorylated cyclic AMP response element binding protein and neuropeptide Y decline as early as middle age in the dentate gyrus and CA1 and CA3 subfields of the hippocampus. Exp Neurol 195: 353–371
21. Shetty AK, Hattiangady B, Shetty GA (2005) Stem/progenitor cell proliferation factors FGF-2, IGF-1, and VEGF exhibit early decline during the course of aging in the hippocampus: role of astrocytes. Glia 51: 173–186
22. Sapolsky RM (1992) Do glucocorticoid concentrations rise with age in the rat? Neurobiol Aging 13: 171–174
23. Herber DL, Mercer M, Roth LM et al. (2007) Microglial activation is required for Abeta clearance after intracranial injection of lipopolysaccharide in APP transgenic mice. J Neuroimmune Pharmacol 2: 222–231
24. Stromberg I, Gemma C, Vila J, Bickford PC (2005) Blueberry- and spirulina-enriched diets enhance striatal dopamine recovery and induce a rapid, transient microglia activation after injury of the rat nigrostriatal dopamine system. Exp Neurol 196: 298–307
25. Hirsch EC, Hunot S, Damier P, Faucheux B (1998) Glial cells and inflammation in Parkinson's disease: a role in neurodegeneration? Ann Neurol 44: S115–S120
26. Gemma C, Catlow B, Cole M et al. (2007) Early inhibition of TNFalpha increases 6-hydroxydopamine-induced striatal degeneration. Brain Res 1147: 240–247

27. Ekdahl CT, Claasen J-H, Bonde S, Kokaia Z, Lindvall O (2003) Inflammation is detrimental for neurogenesis in adult brain. Proc Natl Acad Sci USA 100: 13632–13637
28. Monje ML, Toda H, Palmer TD (2003) Inflammatory blockade restores adult hippocampal neurogenesis. Science 302: 1760–1765
29. Roubenoff R, Harris TB, Abad LW, Wilson PW, Dallal GE, Dinarello CA (1998) Monocyte cytokine production in an elderly population: effect of age and inflammation. J Gerontol A Biol Sci Med Sci 53: M20–M26
30. Wilson CJ, Finch CE, Cohen HJ (2002) Cytokines and cognition – the case for a head-to-toe inflammatory paradigm. J Am Geriatr Soc 50: 2041–2056
31. Gemma C, Mesches MH, Sepesi B, Choo K, Holmes DB, Bickford PC (2002) Diets enriched in foods with high antioxidant activity reverse age-induced decreases in cerebellar beta-adrenergic function and increases in proinflammatory cytokines. J Neurosci 22: 6114–6120
32. Barrientos RM, Higgins EA, Sprunger DB, Watkins LR, Rudy JW, Maier SF (2002) Memory for context is impaired by a post context exposure injection of interleukin-1 beta into dorsal hippocampus. Behav Brain Res 134: 291–298
33. Lynch MA (1998) Analysis of the mechanisms underlying the age-related impairment in long-term potentiation in the rat. Rev Neurosci 9: 169–201
34. Pugh CR, Nguyen KT, Gonyea JL et al. (1999) Role of interleukin-1 beta in impairment of contextual fear conditioning caused by social isolation. Behav Brain Res 106: 109–118
35. Rachal Pugh C, Fleshner M, Watkins LR, Maier SF, Rudy JW (2001) The immune system and memory consolidation: a role for the cytokine IL-1beta. Neurosci Biobehav Rev 25: 29–41
36. Benveniste EN (1992) Inflammatory cytokines within the central nervous system: sources, function, and mechanism of action. Am J Physiol 263: C1–C16
37. Rothwell N, Allan S, Toulmond S (1997) The role of interleukin 1 in acute neurodegeneration and stroke: pathophysiological and therapeutic implications. J Clin Invest 100: 2648–2652
38. Dinarello CA (1998) Interleukin-1, interleukin-1 receptors and interleukin-1 receptor antagonist. Int Rev Immunol 16: 457–499
39. Farrar WL, Kilian PL, Ruff MR, Hill JM, Pert CB (1987) Visualization and characterization of interleukin 1 receptors in brain. J Immunol 139: 459–463
40. Takao T, Tracey DE, Mitchell WM, De Souza EB (1990) Interleukin-1 receptors in mouse brain: characterization and neuronal localization. Endocrinology 127: 3070–3078
41. Sims JE (2002) IL-1 and IL-18 receptors, and their extended family. Curr Opin Immunol 14: 117–122
42. Mesches MH, Gemma C, Veng LM et al. (2004) Sulindac improves memory and increases NMDA receptor subunits in aged Fischer 344 rats. Neurobiol Aging 25: 315–324
43. Gemma C, Fister M, Hudson C, Bickford PC (2005) Improvement of memory for context by inhibition of caspase-1 in aged rats. Eur J Neurosci 22: 1751–1756
44. Gemma C, Bachstetter AD, Cole MJ, Fister M, Hudson C, Bickford PC (2007) Blockade of caspase-1 increases neurogenesis in the aged hippocampus. Eur J Neurosci 26: 2795–2803
45. Wang X, Fu S, Wang Y et al. (2007) Interleukin-1beta mediates proliferation and differentiation of multipotent neural precursor cells through the activation of SAPK/JNK pathway. Mol Cell Neurosci 36: 343–354
46. Iosif RE, Ekdahl CT, Ahlenius H et al. (2006) Tumor necrosis factor receptor 1 is a negative regulator of progenitor proliferation in adult hippocampal neurogenesis. J Neurosci 26: 9703–9712
47. Koo JW, Duman RS (2008) IL-1beta is an essential mediator of the antineurogenic and anhedonic effects of stress. Proc Natl Acad Sci USA 105: 751–756
48. Aloisi F (2001) Immune function of microglia. Glia 36: 165–179
49. Gao HM, Jiang J, Wilson B, Zhang W, Hong JS, Liu B (2002) Microglial activation-mediated delayed and progressive degeneration of rat nigral dopaminergic neurons: relevance to Parkinson's disease. J Neurochem 81: 1285–1297
50. Mantovani A, Sica A, Sozzani S, Allavena P, Vecchi A, Locati M (2004) The chemokine system in diverse forms of macrophage activation and polarization. Trends Immunol 25: 677–686

51. Batchelor PE, Liberatore GT, Wong JY et al. (1999) Activated macrophages and microglia induce dopaminergic sprouting in the injured striatum and express brain-derived neurotrophic factor and glial cell line-derived neurotrophic factor. J Neurosci 19: 1708–1716

52. Miwa T, Furukawa S, Nakajima K, Furukawa Y, Kohsaka S (1997) Lipopolysaccharide enhances synthesis of brain-derived neurotrophic factor in cultured rat microglia. J Neurosci Res 50: 1023–1029

53. Nakajima K, Kohsaka S (2004) Microglia: neuroprotective and neurotrophic cells in the central nervous system. Curr Drug Targets Cardiovasc Haematol Disord 4: 65–84

54. Battista D, Ferrari CC, Gage FH, Pitossi FJ (2006) Neurogenic niche modulation by activated microglia: transforming growth factor beta increases neurogenesis in the adult dentate gyrus. Eur J Neurosci 23: 83–93

55. Butovsky O, Ziv Y, Schwartz A et al. (2006) Microglia activated by IL-4 or IFN-gamma differentially induce neurogenesis and oligodendrogenesis from adult stem/progenitor cells. Mol Cell Neurosci 31: 149–160

56. Schwartz M, Butovsky O, Bruck W, Hanisch UK (2006) Microglial phenotype: is the commitment reversible? Trends Neurosci 29: 68–74

57. Streit WJ (2006) Microglial senescence: does the brain's immune system have an expiration date? Trends Neurosci 29: 506–510

58. Bachstetter AD, Pabon MM, Cole MJ et al. (2008) Peripheral injection of human umbilical cord blood stimulates neurogenesis in the aged rat brain. BMC Neurosci 9: 22

59. Bodles AM, Barger SW (2004) Cytokines and the aging brain – what we don't know might help us. Trends Neurosci 27: 621–626

60. Joseph JA, Shukitt-Hale B, Casadesus G, Fisher D (2005) Oxidative stress and inflammation in brain aging: nutritional considerations. Neurochem Res 30: 927–935

61. Mrak RE, Griffin WS (2005) Glia and their cytokines in progression of neurodegeneration. Neurobiol Aging 26: 349–354

62. Goronzy JJ, Weyand CM (2005) T cell development and receptor diversity during aging. Curr Opin Immunol 17: 468–475

63. Miller RA (1996) The aging immune system: primer and prospectus. Science 273: 70–74

64. Cserr HF, Knopf PM (1992) Cervical lymphatics, the blood-brain barrier and the immunoreactivity of the brain: a new view. Immunol Today 13: 507–512

65. Hickey WF (2001) Basic principles of immunological surveillance of the normal central nervous system. Glia 36: 118–124

66. Kim KD, Zhao J, Auh S et al. (2007) Adaptive immune cells temper initial innate responses. Nat Med 13: 1248–1252

67. Stichel CC, Luebbert H (2007) Inflammatory processes in the aging mouse brain: participation of dendritic cells and T-cells. Neurobiol Aging 28: 1507–1521

68. Bulloch K, Miller MM, Gal-Toth J et al. (2008) CD11c/EYFP transgene illuminates a discrete network of dendritic cells within the embryonic, neonatal, adult, and injured mouse brain. J Comp Neurol 508: 687–710

69. Mooradian AD (1988) Effect of aging on the blood–brain barrier. Neurobiol Aging 9: 31–39

70. Mooradian AD (1994) Potential mechanisms of the age-related changes in the blood–brain barrier. Neurobiol Aging 15: 751–755

71. Morita T, Mizutani Y, Sawada M, Shimada A (2005) Immunohistochemical and ultrastructural findings related to the blood–brain barrier in the blood vessels of the cerebral white matter in aged dogs. J Comp Pathol 133: 14–22

72. Kipnis J, Avidan H, Caspi RR, Schwartz M (2004) Dual effect of CD4+CD25+ regulatory T cells in neurodegeneration: a dialogue with microglia. Proc Natl Acad Sci USA 101 Suppl 2: 14663–14669

73. Butovsky O, Koronyo-Hamaoui M, Kunis G et al. (2006) Glatiramer acetate fights against Alzheimer's disease by inducing dendritic-like microglia expressing insulin-like growth factor 1. Proc Natl Acad Sci USA 103: 11784–11789

Neuronal Cytoskeleton and the Tau Hypothesis

Tau Transgenic Mouse Models in Therapeutic Development

Hanno M. Roder

Abstract Although a unifying animal model of Alzheimer's disease (AD) has not been forthcoming, specific pathological features have been successfully induced in transgenic mice with mutated human genes identified by genetic analysis. A particular problem is that the defining pathologies of AD, amyloid-β peptide amyloidosis, and neurofibrillary tangles formed by the abnormally modified microtubule-associated protein tau cannot both be induced by a single mutation as in humans, calling in question straightforward cause–effect hypotheses. On the contrary, the separate manifestation of these pathologies in mice points to their distinct functional consequences: amyloid-β pathology interferes with synaptic efficiency but does not by itself drive neurodegeneration, while tau pathology appears to be the direct mechanism of neuronal degeneration and brain atrophy. In both cases, however, the latest data favor a dominant role of biochemical precursors to the obvious neuropathological structures, which originally defined the disease. In any case, for therapeutic development, tau pathology needs to be addressed for lasting treatment results in AD. Initial therapeutic studies with mouse models of tauopathy support their utility in the discovery of antineurodegenerative therapeutic principles. Moreover, certain aspects of tau pathology and their functional consequences may be partially reversible. Apart from such immediate utility, transgenic mouse models also promise to provide unique insights into the biochemical details of tau pathology, which are impossible to obtain from human AD brains.

1 Challenges to Modeling Alzheimer's Disease

The understanding of the molecular pathology of Alzheimer's disease (AD) and with it the rational conception of prospective therapeutic ideas are still obstructed by the lack of a unifying concept withstanding all experimental tests. The difficulty

H.M. Roder
Tau Ta Tis, Inc.
P.O. Box 350849, Jacksonville, FL 32235

R.B. Maccioni and G. Perry (eds.) *Current Hypotheses and Research Milestones in Alzheimer's Disease*. DOI: 10.1007/978-0-387-87995-6_4,
© Springer Science + Business Media, LLC 2009

is related to a schisma set up in the original definition of the disease as formulated by Alois Alzheimer a century ago. Descriptive studies of the main pathological lesions, extracellular aggregates of amyloid-β (Aβ) peptides derived from a transmembrane amyloid-β protein precursor (AβPP), and intracellular aggregates of the abnormally modified cytoskeletal protein tau have not resolved cause–effect issues to a detail as desirable for systematic drug discovery. One of the problems recognized early is that the two lesions do not have to occur together. Most important, the presence of parenchymal Aβ deposits at high levels, in the absence of vascular involvement, is often not associated with clinical dementia [1–4], while pure tauopathies (some types of frontotemporal dementia, Pick's disease, corticobasal degeneration, etc.) are without exception associated with the functional deficits corresponding to the affected brain areas. Moreover, rates of brain atrophy are not related to any measure of pathologically assessable Aβ deposition [5]. Differences in the nature of the amyloid depositions between normal and AD subjects could be invoked to account for this discrepancy [6, 7]; however, it is unclear how this would interface with an increasingly discussed role of soluble Aβ upstream of deposition in neurodegeneration (see below). Not necessarily functionally committed to Aβ pathology, genetic analyses of relatively rare early onset AD have provided clear evidence for the ability of AβPP-related biology to cause the disease [8, 9] in conjunction with generic age-related changes, which remain anonymous, however. This fact has fostered the development of a fairly broad consensus that the key to truly fundamental and therefore curative therapeutic interventions must lie within the biochemistry of AβPP, and more specifically, its proteolytic processing to produce the aggregating Aβ peptides [10]. Various alterations in processing linked to pathogenic mutations in AβPP and to mutations in two of its proteases (presenilins, γ-secretase) have contributed to this view (e.g., [11–13]).

Although the correlative evidence amassed for a central position of Aβ in therapeutic concepts is intriguing, the advent of animal models based on human pathogenic mutations in key AD molecules has given further cause to believe that something must be amiss in a sequence of events termed the "Amyloid Cascade Hypothesis" [10]. AβPP or AβPP/PS1 transgenic mice with human disease-causing mutations indeed produce massive amyloidoses, but no neurodegeneration [14, 15]. There are, however, functional deficits in learning and memory, and those may be related to impairment of synaptic function by soluble Aβ species, rather than plaques [16, 17]. These observations have directed attention to the pathological activity of precursors of amyloid plaques, and their acute role in such mouse models is supported by the nearly immediate and dramatic reversal of deficits after passive immunization [18, 19]. However, such reversals have not been seen in a pioneering clinical trial of Aβ vaccination in AD patients [20], although postmortem examination showed that the pathological objective, that is, substantial clearance of Aβ-plaques, was indeed achieved [21]. Even a slow-down of deterioration of treated over placebo AD groups could not be clearly demonstrated after long-term follow-up of those patients with generally maintained anti-Aβ titers, which had escaped the side effects [20]. This disconnect invites the interpretation that the functional effects of soluble Aβ-species in mice are dramatically exaggerated by the overexpression of mutant AβPP (and thereby Aβ), which is required to precipitate Aβ-pathology in such mice, and is sufficiently low in human AD patients to be difficult to detect on the background of more dominating events. The mouse data further suggest that even if pathological Aβ-species are cleared, the therapeutic effect in the best case may not amount to much more than an offset of acute impairment, very much like cognition enhancers.

Notwithstanding a primary causal role of AβPP-biology, the more dominating contributor to functional impairment in AD patients with active disease may therefore be the substantially irreversible neurodegenerative activity rather than the component of acute synaptic impairment. The synopsis of the above body of findings suggests that consistently intracellular non-Aβ pathologies, like Lewy bodies, vascular pathologies, or more frequently tauopathy in the case of classical AD, may be driving the functionally dominating aspect of clinical AD, that is, the loss of brain mass [5]. The failure of mouse models of Aβ-pathology to recapitulate the tauopathy aspect could be blamed on endogenous mouse tau, which may simply lack some biochemical prerequisite to transduce the pathological impetus of the "amyloid cascade" in the way the human homologue does [22]. Yet wild-type human tau in the mouse context is also not coerced into PHF (Paired Helical Filament) pathology on the background of AβPP/PS1 transgenic mouse pathology [23]. Only mice carrying pathogenic mutant human tau, causing "tau-only" pathology in both mice and humans, respond with enhanced tau pathology in the presence of Aβ-pathology [24–26]. However, in the absence of a clear de novo induction of tauopathy in an animal model, corresponding to the signal-transduction nature of the "amyloid cascade" proposition, there is no way to ascertain whether this synergy represents a specific interaction of the respective pathologies as entertained for AD, or to even exclude an artifactual result.

On this background, tau transgenic mice with the bona fide tau pathology of AD emerge as indispensable tools for therapeutic AD research.

2   Models of the Tauopathy Aspect of AD

Neurofibrillary pathology, similar in structural and biochemical aspects, is also found in a variety of less common neurological conditions other than AD, as defined by the absence of concomitant Aβ-plaque pathology. The symptomatology corresponds to functional involvement of the brain regions afflicted (e.g., an initially more psychiatric phenotype in FTDP-17). The progressive nature of all of these conditions in the absence of other pathologies emphasizes the self-propagating nature of tau pathology on its own, and disease durations are not unlike those of AD. As a consequence, it is highly plausible that tauopathies may progress independently of their cause, once started. For AD, it raises the specter that in spite of AβPP-biology being able to constitute a single sufficient cause of AD, its antagonism may not be fully effective with tauopathy underway [27]. Tauopathy is already manifest in mild-cognitive-impairment patients [7], which convert with high likelihood to AD [28].

In some instances, tauopathies have an autosomal-dominant inheritance pattern, and this was traced to mutations in tau [29]. Transgenic mice producing the same mutant human tau proteins develop neurofibrillary pathology very similar to AD [30], accompanied by unambiguous neurodegeneration [31]. More specifically, the pathological tau protein in such mice is found in a qualitatively abnormal form, indistinguishable from the PHF-tau proteins found in AD and all other tauopathies. The conversion of normal tau into PHF-tau is effected by abnormal hyperphosphorylation, which is unable to perform its normal function related to binding to microtubules, and aggregates. All known markers of PHF-type hyperphosphorylation are reflected in the hyperphosphorylated human tau expressed in transgenic mice [32], including an abnormal SDS-resistant conformation reported by a uniquely retarded SDS-PAGE mobility behavior, which is never seen in any physiological situation. There appears to be no normal context for tau hyperphosphorylation of this type. This is decisively distinct from many other biochemical abnormalities in disease biology, which are quantitative deviations: increased Aβ levels, over- or underactivation of pathways in proliferation and inflammation, abnormal levels of factors and hormones, etc. The faithful recapitulation of PHF-tau pathology in tau transgenic mice makes them attractive models to test specific pathomechanism-directed anti-neurodegenerative therapeutic strategies.

3   Tau Hyperphosphorylation as a Key Therapeutic Target

The characteristic qualitatively abnormal hyperphosphorylation of tau is the common denominator of all tauopathies and AD [33]. In transgenic mice, the onset of pathology and functional impairment is not seen prior to the emergence of this biochemical hallmark. This suggests that with or without mutations of tau, it is the hyperphosphorylation which converts a latent problem into active pathology during aging. Inhibition of this process is therefore highly plausible to prevent, retard, or possibly even partially reverse the pathology in a causal way.

Although transgenic models of tauopathy based on mutations of tau have been available since the year 2000, only few reports of their use for therapeutic test purposes have surfaced to date (Table 1). Addressing a surmised abnormal conformation of mutant tau, hsp90 inhibitors have been tested in the JNPL3 mouse model with spinal cord pathology and motor phenotype [34]. In analogy to mutant proteins in cancer, inhibition of heat-shock protein (hsp)90 (i.e., its ATPase activity) leads to reduced chaperone activity and enhanced degradation of the mutant tau protein over normal tau protein. This led to reduced formation of sarcosyl-insoluble PHF-tau species, which are frequently used as a biochemical proxy to pathologically defined neurofibrillary tangles. However, the relationship of these operationally defined species to the overall scope of tau hyperphosphorylation is less certain than generally assumed (*vide infra*). Moreover, no impact on phenotype was described. It is also unclear whether this approach would be viable in the much more frequent tauopathies not driven by mutations in tau, where tau levels might be unresponsive to hsp90 inhibition. In view of this concern, mutant tau transgenic models may not be particularly predictive for AD for this specific approach. On the other hand, the levels of a host of other proteins not involved in pathology are likely to be affected in view of the normal function of hsp90.

Other studies using mutant tau transgenic mice sought to directly address inhibition of tau hyperphosphorylation. Since GSK3-β has been implicated as a tau kinase, JNPL3 mice were treated with therapeutic levels of lithium, a specific GSK3 inhibitor [35]. There was a reduction in pathological markers when applied before the age of onset of symptoms, but it is questionable whether this was related to GSK3 inhibition in view of the increasing recognition of lithium as an enhancer of autophagy via inositol–monophosphatase inhibition [36]. In any case, a retardation of functional impairment was not seen with this treatment, which corroborates results discussed below that the generally used pathological markers of tauopathy, which biochemically represent downstream products, may not be the dominant toxic tau species.

Significant efficacy, even at ages around disease onset, on the level of abnormally hyperphosphorylated tau species as well as for onset and/or progression of motor impairment in the same model was observed with the orally bioavailable and brain-penetrating compound SRN-003-556, exhibiting a broader kinase inhibitory spectrum [37]. This compound was developed on the basis of efficacy in cell models of tau hyperphosphorylation induced by the mainly PP2A-inhibiting agent okadaic acid [38], which are the only models where all criteria of PHF-type tau hyperphosphorylation (including loss of microtubule-binding) become manifest. Only inhibitors including the inhibition of ERK2 in their spectrum were active in this model, although inhibition of ERK2 alone was not sufficient. Consequently, SRN-003-556 was optimized with ERK2 as a lead assay, although collateral inhibitory activities are probably relevant as well. A surprising observation in this study was the apparent uncoupling of tau hyperphosphorylation, determined by biochemical means, from pathologically assessed neurofibrillary tangle counts. While in the absence of treatment, those measures correlated well, as expected, under treatment, only the amount of hyperphosphorylated tau was

Table 1 Therapeutic approaches for tau pathology tested in mutant tau transgenic mice

Target	Mouse model	Agent	PHF-tau reduction	Tangle reduction	Phenotype improvement	Reference
GSK3	JNPL3	Lithium	?	Yes	No	[35]
Cdk5	Not tested	–	–	–	–	–
MAP-kinases	JNPL3	SRN-003-556	Yes	No	Yes	[37]
HSP-90	JNPL3	Small molecule	Yes	?	?	[34]

reduced significantly, but not the tangle counts [37]. Hyperphosphorylated tau was determined directly from whole brain supernatants, and not using the proxy of sarcosyl-insoluble PHF-tau obtained through further fractionation of the former. It has since been noted that sarcosyl-insoluble PHF-tau represents only a rather marginal fraction of the total pool of hyperphosphorylated tau [39], which in some cases can even exceed the levels of normal unaffected tau. Moreover, the sarcosyl-insoluble tau pool from the JNPL3 inhibitor study remained also unaffected by treatment (H. Roder, unpublished observations), indicating a closer relationship with the tangle pool, and calling into question the use of that pool to assess tau kinase pharmacology. A possible explanation for this curious discrepancy might be that formation of histopathologically defined tangles is a much slower (rate-limiting) process than biochemical formation of abnormally hyperphosphorylated tau. Such a partial uncoupling between upstream pathological tau biochemistry and downstream classical markers of neurofibrillary degeneration could potentially explain findings of neurodegeneration in tau-transgenic mouse models seemingly independent of classical neuropathological tau markers [40, 41]. However, this would also imply a lower toxicity of tangles compared to its precursors.

These results match those obtained by reducing tau pathology using genetic means. In a conditional transgenic mouse model of cortical tau pathology with a learning and memory phenotype, suppression of mutant tau expression after onset of tau pathology did not reduce sarcosyl-insoluble tau species or tangle counts, but halted neurodegeneration and led even to a moderate recovery of memory function [42]. It is therefore conceivable that upstream aspects of tau pathology might be reversible to some degree, and that the respective therapeutic principles could not only act as mere modifiers of progression but also lead to immediate relaxation of neurological symptoms. This prospect is yet to be confirmed by pharmacological means.

4 Conclusion

Tau transgenic mice have become indispensable tools in AD-therapeutic research. They model all known features of the pathology: the highly specific type of hyperphosphorylation of tau, the aggregation of this abnormal tau into straight or paired helical filaments, neurofibrillary tangles with the full complement of antigenic properties, and robust neuritic and cellular degeneration with commensurate functional deficits. Although the entire chain of events of AD is certainly not modeled in such mice, all processes at the level or downstream of PHF-tau formation, such as hyperphosphorylation, aggregation, clearance, loss of neurons, and functional impairment are very accessible in reasonable time frames, and provide highly credible readouts to test any therapeutic intervention strategy. Efficacy in such models, especially in view of their aggressiveness, can be expected to predict efficacy in man.

References

1. Dickson DW, Crystal HA, Mattiace LA et al. (1992) Identification of normal and pathological aging in prospectively studied non-demented elderly humans. Neurobiol Aging 13:179–189
2. Armstrong RA (1994) Beta-amyloid (Abeta) deposition in elderly non-demented patients and patients with Alzheimer's disease. Neurosci Lett 178:59–62
3. Davis DG, Schmitt FA, Wekstein DR, Markesberry WR (1999) Alzheimer neuropathologic alterations in aged cognitively normal subjects. J Neuropathol Exp Neurol 58:376–388
4. Knopman DS, Parisi JE, Salviati A et al. (2003) Neuropathology of cognitively normal elderly. J Neuropathol Exp Neurol 62:1087–1095
5. Josephs KA, Whitwell JL, Ahmed Z et al. (2008) Beta-amyloid burden is not associated with rates of brain atrophy. Ann Neurol 63:204–212
6. Piccini A, Russo C, Gliozzi A et al. (2005) β-Amyloid is different in normal aging and in Alzheimer's disease. J Biol Chem 280:34186–34192
7. Markesbery WR, Schmitt FA, Kryscio RJ et al. (2006) Neuropathologic substrate of mild cognitive impairment. Arch Neurol 63:38–46
8. Goate A, Chartier-Harlin MC, Mullan M et al. (1991) Segregation of a missense mutation in the amyloid precursor protein gene with familial Alzheimer's disease. Nature 349:704–706
9. Lannfelt L, Bogdanovic N, Appelgren H et al. (1994) Amyloid precursor protein mutation causes Alzheimer's disease in a Swedish family. Neurosci Lett 168:254–256
10. Hardy J, Selkoe DJ (2002) The amyloid hypothesis of Alzheimer's disease: progress and problems on the road to therapeutics. Science 297:353–356
11. Haass C, Lemere CA, Capell A et al. (1995) The Swedish mutation causes early-onset Alzheimer's disease by beta-secretase cleavage within the secretory pathway. Nat Med 1:1291–1296
12. Mehta ND, Refolo LM, Eckman C et al. (1998) Increased Abeta42(43) from cell lines expressing presenilin 1 mutations. Ann Neurol 43:256–258
13. Walker ES, Martinez M, Brunkan AL, Goate A (2005) Presenilin 2 familial Alzheimer's disease mutations result in partial loss of function and dramatic changes in Abeta 42/40 ratios. J Neurochem 92:294–301
14. Irizarry MC, Soriano F, McNamara M et al. (1997) Abeta deposition is associated with neuropil changes, but not with overt neuronal loss in the human amyloid precursor protein V717F (PDAPP) transgenic mouse. J Neurosci 17:7053–7059
15. Takeuchi A, Irizarry MC, Duff K et al. (2000) Age-related amyloid beta deposition in transgenic mice overexpressing both Alzheimer mutant presenilin 1 and amyloid beta precursor protein Swedish mutant is not associated with global neuronal loss. Am J Pathol 157:331–339
16. Lesne S, Koh MT, Kotilinek L et al. (2006) A specific amyloid-beta protein assembly in the brain impairs memory. Nature 440:352–357
17. Lesne S, Kotilinek L, Ashe KH (2008) Plaque-bearing mice with reduced levels of oligomeric amyloid-beta assemblies have intact memory function. Neuroscience 151:745–749
18. Dodart JC, Bales KR, Gannon KS et al. (2002) Immunization reverses memory deficits without reducing brain Abeta burden in Alzheimer's disease model. Nat Neurosci 5:452–457
19. Kotilinek LA, Bacskai B, Westerman M et al. (2002) Reversible memory loss in a mouse transgenic model of Alzheimer's disease. J Neurosci 22:6331–6335
20. Gilman S, Koller M, Black RS et al. (2005) Clinical effects of Aβ immunization (AN1792) in patients with AD in an interrupted trial. Neurology 64:1553–1562
21. Nicoll JA, Wilkinson D, Holmes C et al. (2003) Neuropathology of human Alzheimer disease after immunization with amyloid-beta peptide: a case report. Nat Med 9:448–452
22. Xu G, Gonzales V, Borchelt DR (2002) Abeta deposition does not cause the aggregation of endogenous tau in transgenic mice. Alzheimer Dis Assoc Disord 16:196–201
23. Boutajangout A, Authelet M, Blanchard V et al. (2004) Characterisation of cytoskeletal abnormalities in mice transgenic for wild-type human tau and familial Alzheimer's disease mutants of APP and presenilin-1. Neurobiol Dis 15:47–60

24. Lewis J, Dickson DW, Lin WL et al. (2001) Enhanced neurofibrillary degeneration in transgenic mice expressing mutant tau and APP. Science 293:1487–1491
25. Goetz J, Chen F, van Dorpe J, Nitsch RM (2001) Formation of neurofibrillary tangles in P301l tau transgenic mice induced by Abeta 42 fibrils. Science 293:1491–1495
26. Bolmont T, Clavaguera F, Meyer-Luehmann M et al. (2007) Induction of tau pathology by intracerebral infusion of amyloid-beta containing brain extract and by amyloid deposition in APP x tau transgenic mice. Am J Pathol 171:2012–2020
27. Kulic L, Kurosinski P, Chen F et al. (2006) Active immunization trial in Abeta42-injected P301L tau transgenic mice. Neurobiol Dis 22:50–56
28. Jicha GA, Parisi JE, Dickson DW et al. (2006) Neuropathologic outcome of mild cognitive impairment following progression to clinical dementia. Arch Neurol 63:674–681
29. Hutton M, Lendon CL, Rizzu P et al. (1998) Association of missense and 5′-splice-site mutations in tau with the inherited dementia FTDP-17. Nature 393:702–705
30. Lewis J, McGowan E, Rockwood J et al. (2000) Neurofibrillary tangles, amyotrophy and progressive motor disturbance in mice expressing mutant (P301L) tau protein. Nat Genet 25:402–405
31. Ramsden M, Kotilinek L, Forster C et al. (2005) Age-dependent neurofibrillary tangle formation, neuron loss, and memory impairment in a mouse model of human tauopathy (P301L). J Neurosci 25:10637–10647
32. Sahara N, Lewis J, DeTure M et al. (2002) Assembly of tau in transgenic animals expressing P301L tau: alteration of phosphorylation and solubility. J Neurochem 83:1498–1508
33. Mailliot C, Sergeant N, Bussiere T et al. (1998) Phosphorylation of specific sets of tau isoforms reflects different neurofibrillary degeneration processes. FEBS Lett 433:201–204
34. Luo W, Dou F, Rodina A et al. (2007) Roles of heat-shock protein 90 in maintaining and facilitating the neurodegenerative phenotype in tauopathies. Proc Natl Acad Sci USA 104:9511–9516
35. Noble W, Planel E, Zehr C et al. (2005) Inhibition of glycogen synthase kinase-3 by lithium correlates with reduced tauopathy and degeneration in vivo. Proc Natl Acad Sci USA 102:6990–6995
36. Sakar S, Floto RA, Berger Z et al. (2005) Lithium induces autophagy by inhibiting inositol monophosphatase. J Cell Biol 170:1101–1111
37. LeCorre S, Klafki HW, Plesnila N et al. (2006) An inhibitor of tau hyperphosphorylation prevents severe motor impairments in tau transgenic mice. Proc Natl Acad Sci USA 103:9673–9678
38. Huebinger G, Geis S, LeCorre S et al. (2008) Inhibition of PHF-like tau hyperphosphorylation in SH-SY5Y cells and rat brain slices by K252a. J Alzheimers Dis 13(3):281–294
39. Berger Z, Roder H, Hanna A et al. (2007) Accumulation of pathological tau species and memory loss in a conditional model of tauopathy. J Neurosci 27:3650–3662
40. Spires TL, Orne JD, Santacruz K et al. (2006) Region-specific dissociation of neuronal loss and neurofibrillary pathology in a mouse model of tauopathy. Am J Pathol 168:1598–1607
41. Andorfer C, Kress Y, Espinoza M et al. (2003) Hyperphosphorylation and aggregation of tau in mice expressing normal human tau isoforms. J Neurochem 86:582–560
42. Santacruz K, Lewis J, Spires T et al. (2005) Tau suppression in a neurodegenerative mouse model improves memory function. Science 309:476–481

What Have We Learned from the Tau Hypothesis?

Ricardo B. Maccioni, Gustavo A. Farias, Leonel E. Rojo, M. Alejandra Sekler, and Rodrigo O. Kuljis

Abstract Until the appearance of the Danger Signals Hypothesis on Alzheimer's disease (AD), none of the hypotheses on its pathogenesis accounted coherently for the diversity of the earliest events that trigger neurodegeneration, and that eventually result in senile plaques (SP) and neurofibrillary tangles (NFTs). The original version of the most commonly held amyloid hypothesis rests on the concept that amyloid-β $(A\beta)_{1-42}$ self-polymerizes over many years to form SP, which then triggers the entire array of subsequent brain lesions. However, recent findings point to unpleated $A\beta$ oligomers (ADDLs) as the major culprit for synaptic impairment, well before neuronal degeneration ensues. Amyloid deposits thus appear to be a rather late event in a long chain driving progressively more severe neuronal, glial and neuropil alterations. AD is a multifactorial disorder in that protein alterations, oxidative stress, neuroinflammation, immune deregulation, impairment of neuronal-glial communication, and neurotoxic agents appear to be the major factors triggering neuronal degeneration, and the balance among these seems to vary from patient to patient. Although diverse, these factors induce deleterious signaling through different sets of neuronal receptors that converge in the hyperphosphorylation of tau molecules. Thus, tau hyperphosphorylation constitutes a common final pathway for most of the altered molecular and cellular factors that eventually result in degenerating neurons. This raises the question as to precisely what triggers the pathological phosphorylation. We have shown that $A\beta$ oligomers, oxygen free radicals, and iron overload destabilize the equilibrium between the activities of protein phosphatases and kinases involved in tau assembly. Furthermore, overproduction or processing alterations of trophic factors such as NGF by activated glial cells trigger signaling cascades via p75, leading to cdk5 activation, followed by tau hyperphosphorylation and neuronal death. The cytokines TNFα, IL-1, and IL-6 induce activation of the cdk5/p35 complex, which causes tau phosphorylation.

R.B. Maccioni(✉), G.A. Farias, L.E. Rojo, M.A. Sekler, and R.O. Kuljis
Department of Neurology
Faculty of Medicine & Lab. Cell Mol. Neurosci.
Faculty Sciences, U Chile, Las Encinas 3370,
Ñuñoa, Santiago, Chile
e-mail: rmaccion@manquehue.net

R.B. Maccioni and G. Perry (eds.) *Current Hypotheses and Research Milestones in Alzheimer's Disease*. DOI: 10.1007/978-0-387-87995-6_5,
© Springer Science + Business Media, LLC 2009

49

Converging lines of evidence reveal the involvement of innate immunity (in contrast with the more widely acknowledged, but probably less-important involvement of adaptive immunity) and the role of inflammatory processes in the development of AD-associated neuronal changes. While methodological challenges cannot be ruled out in the interpretation of the so far confusing and sometimes even contradictory clinical trials, inflammation is essential in virtually all animal models for AD-like lesions. Taken together, these observations indicate that slowly accumulating danger/alarm signals to the innate immune system interfere with the balance of protective versus degeneration-promoting mechanisms, shifting the equilibrium toward neurodegeneration that involves deregulation of protein kinases cdk5 and GSKβ, tau hyperphosphorylation, and its aggregation into anomalous polymers in the neuronal cytoskeleton that constitutes the converging result of a large array of risk factors over time. These mediate the inexorable worsening of cognitive manifestations along with neuronal degeneration and the eventual appearance of tardy lesions such as SP and NFTs. This new theoretical framework based on recent experimental findings may serve as a powerful tool in the development of the much-sought biomarkers and in vivo imaging technology for the early diagnosis of AD. This will also help in the design of effective interventions to both treat and perhaps even prevent this increasingly prevalent brain disorder.

Abbreviations Aβ, amyloid-β; AD, Alzheimer's disease; ADDLs, Aβ oligomers neurotoxic for the synapses, β-structure, secondary structure type beta of a protein; Cdk5, cyclin-dependent protein kinase 5; C-terminal domain, carboxyl-terminal domain, GSK3β, glycogen synthase kinase β; CSF, cerebrospinal fluid; IL-1 and -6, interleukins 1 and 6, LDL, low density lipoproteins, MAPs, microtubule-associated proteins; NFTs, neurofibrillary tangles; PHFs, paired helical filaments; SP, senile plaques; TNFα, tumor necrosis factor-α.

1   Introduction

The discovery by Alois Alzheimer of neurofibrillary tangles (NFTs) in the brains of patients with the neurodegenerative disorder named after him [1] provided a pivotal impetus for the study of their molecular substrate [2, 3]. However, it was only in the 1980s that it became evident that the major components of NFTs, the paired helical filaments (PHFs), are mainly formed by a hyperphosphorylated form of the tau protein [4, 5]. Many studies have since improved our understanding of tau hyperphosphorylation, changes in tau interaction patterns with microtubules, and alteration of the neuronal cytoskeleton [3, 6], but the structural basis for tau self-aggregation is still a major scientific puzzle that remains to be elucidated. On the basis of structural studies, together with the elucidation of the signaling cascades in neurodegeneration, we postulated a hypothesis based on the concept that *tau hyperphosphorylation constitutes a final common pathway in the pathogenesis of Alzheimer's disease (AD)*, upon which a host of signaling mechanisms converge, and that this phenomenon precedes widespread neuronal degeneration [7–17].

Thus, the tau hypothesis is based on a host of converging studies [2, 18–21] and is revisited here in the light of recent findings and the increasingly evident role of altered immunomodulation in AD [22, 23]. We believe that the establishment of an experimentally testable unifying hypothesis on AD is the best basis on which to build innovative diagnostic and therapeutic approaches. The virtually unchallenged acceptance of the amyloid hypothesis for around two decades, after Glenner and Wong [24] postulated it, has led to unsuccessful and costly efforts to generate drugs to control AD, since pleated amyloid and senile plaques (SP) are most probably not the cause of neuronal degeneration [23]. Thus, the lofty expectations harbored on antiamyloid therapies will most likely not materialize. In fact, the revised version of the most commonly held amyloid hypothesis [25] rests on the concept that amyloid-β $(A\beta)_{1-42}$ self-polymerizes over years to form SP, which then trigger the entire array of subsequent brain lesions. However, increasing numbers of recent findings point to unpleated $A\beta$ oligomers (ADDLs) as the elements responsible for synaptic impairment well before the "classical" lesions develop, including accumulations of fibrillary amyloid [26]. Moreover, SPs are also common in neurologically healthy individuals, and not unique to AD. On the basis of these and additional evidence, followers of the tau hypothesis have pointed to the fact that tau aggregation and neurotoxicity associated with the hyperphosphorylated forms constitute common events determinant for the neurodegenerative cascade, and that amyloid pathology could be one among many other risk factors for the disease.

Recent evidence reviewed here indicates increasingly and unequivocally that AD is a multifactorial disorder [2, 21] and consequently, hypotheses on the etiology of AD that account for this fact are more likely to be experimentally corroborated and lead to successful therapeutic and preventative interventions. Therefore, we focus our discussion on the updated tau hypothesis together with current neuroimmunomodulation concepts that provide a novel unifying hypothesis of AD etiology and pathophysiology that account coherently for virtually all known facts of the disorder and point the way for subsequent efforts at understanding and conquering it. Therefore, the information obtained in the process of testing this new hypothesis experimentally will likely be helpful to formulate an innovative AD therapy and to design reliable biomarker strategies for its diagnosis [27–29].

2 Alzheimer's Disease: A Pivotal Focus for Innovative Biomedical Technology Development

The classical era of neurological sciences during the end of the nineteenth century and the last century was characterized by pivotal discoveries such as synaptic transmission, the complexity of brain functions, and the correlation between neuroanatomical information and neurophysiology from Cajal to the present [30, 31]. With the rise of modern cell biology and molecular genetics four decades ago, a broad area of innovative development began, especially in the Neurosciences, that, among other subjects, promises to provide solutions to the challenges of prevention, diagnosis, and treatment of neurodegenerative diseases. Besides the strictly medical area, the Neurosciences integrate domains such as neuropsychology and linguistics that serve as a bridge between "hard core" sciences such as biology and physics, with the humanities [32]. Pathologies that affect cognitive processes of the brain include three major groups of diseases: neurodegenerative disorders, with AD being by far the most prevalent disease – and the one with a dramatically increasing incidence – stroke, and brain trauma. The social and economical cost to public health of all these pathologies is enormous. Therefore, scientific endeavors through decades have led to important findings on the alterations in the proteins involved in neurodegenerative disorders, on the structure and cellular functions of tau protein, and on the relationships that link the effects of the amyloid-β and tau hyperphosphorylations in the pathogenesis of AD.

AD is the principal cause of dementia throughout the world, and the fourth cause of death in well-developed economies after cancer, cardiovascular diseases, and stroke. However, the set of disorders that causes cognitive impairment, which includes vascular brain disorders and head injury, is the largest cause of morbidity and mortality. More than 5 million people are affected by this disease in USA alone, and mortality is over 100,000 per year, with a cost to the economy that exceeds US\$ 100 billion [33–36]. Thus, AD constitutes nowadays one of the largest health problems in the world, since projections for year 2010 indicate that more than 36 million people will have AD.

As mentioned previously, it is clear that AD is a disease caused by multiple factors – and not a single unifying "cause" – that besides tau aggregates involve, among many signals triggering the neurodegenerative cascade, the presence of soluble neurotoxic amyloid oligomers (ADDLs), neuroinflammatory cytokines, deregulation of neuron-glial cells, cross talks mediated by changes in the production and processing of growth factors, proinflammatory cytokines and other signaling molecules, and oxidative stress [23]. Despite the apparent absence of tau mutations in familial forms of AD, the tau hypothesis has increasingly stronger support from studies in cells and animal models and also from the clinical and neuropathological evidence indicating that tau pathology is pathognomonic of AD. Furthermore, in tauopathies associated with frontotemporal dementia (FTD-17), point mutations on tau are clearly an established cause of dementia [37], demonstrating that tau alterations alone are sufficient to produce cognitive deterioration independently of the risk factors that lead to AD. Therefore, the tau hypothesis of AD is indispensable among the efforts to establish a bridge between basic research discoveries on tau and its biomedical implications.

3 Tau and Tau Aggregates

The low molecular weight microtubule-associated protein (MAP) tau is the major component of the MAPs in axons, and plays critical physiological roles in stabilizing microtubules and inducing its own assembly [6, 38]. One of the most intriguing properties of this brain polypeptide is that, under pathological conditions, tau self-aggregates into PHFs, which turn into the NFTs during the course of AD, one of the neuropathological hallmarks of AD and tauopathies [3, 39, 40]. However, today we know that hyperphosphorylated tau or oligomers of this protein exert the pathological effects, triggering neurotoxic action and altering the normal interaction patterns of the neuronal cytoskeletal network [41]. Moreover, a link between pathological tau oligomerization and cognitive impairment has been shown [42]. Important advances toward our understanding on in vitro tau polymerization has been provided by a number of authors, including those from our laboratory [4, 12, 13, 43–45]. However, the mechanisms underlying the structural transition from an innocuous, natively unfolded, tau to its neurotoxic polymers remain unknown, as is the detailed structural mechanisms of this macromolecular aggregation.

Tau protein binds to microtubules through the "repeat domain" in the C-terminal fragment, two flanking regions are also important for the process of microtubule polymerization [8, 46, 47]. Simulation models in addition to binding analysis have accounted for the nature of tau association with microtubules [48]. Tau interaction with other cytoskeletal structures has been also shown [49, 50]. Tau variants have been found in centrosomes [51] as well as in the neuronal nucleus, the role of the latter analyzed by Sjoberg et al. [20]. It has been demonstrated that there are two motifs in the upstream flanking domain, ^{225}KVAVVRT^{231} and ^{243}LQTA^{246}, and one downstream of the repeats, ^{370}KIETHKTFREN^{380}, which strongly contribute to the binding to the acidic outside of microtubules as well as to the binding of other polyanions such as heparin, which is a process commonly used to form PHFs in vitro [44, 52]. Advances have been obtained on tau arrangement within PHFs, being present in discrete nonconfluent patterns within PHFs forming bridges that interconnect individual PHFs to form the complex macromolecular network known as NFTs [53]. Tau aggregation results from transitions from random coiled domains to β-structure [54]. NMR studies allowed the characterization of a 198-residue tau fragment composed of the 4 tandem repeats and the flanking domains and containing the full microtubule binding and tau assembling activity. The highest propensity for β;-structure is within the four-repeat region, whereas the flanking domains are largely random coil, with an increased rigidity in the proline-rich region [55].

Chemical shift perturbation studies show that polyanions that promote PHFs aggregation as well as microtubules interact with tau through positive charges near the ends of the repeats and through the β-forming motifs at the beginning of repeats 2 and 3. The high degree of similarity between the binding of polyanions and microtubules supports the hypothesis that stable microtubules prevent PHF formation by blocking the tau-polyanion interaction sites, which are crucial for PHF formation [55]. On the other hand, interesting results point to the role of tau oligomers in early stages of AD and tauopathies. Maeda et al. [56] have identified a granular tau oligomer with a prefilamentous structure which is present in samples of the frontal cortex of asymptomatic patients displaying Braak-stage I neuropathology, a stage where clinical symptoms of AD and NFTs in frontal cortex are believed to be absent. This suggests that the increase in tau oligomer levels occurs before NFTs formation and before individuals manifest clinical symptoms of AD. In this context, the search of ligand with high affinity for tau oligomers may provide a promising avenue for future studies on PET radiotracer for neuroimaging of tau pathology [57].

4   Cellular Cascades Leading to Tau Hyperphosphorylations

Research on the physiopathology of the major molecular factors triggering AD pathogenesis has provided clues on the structural–functional underpinnings of tau–tau interactions [6, 9, 22, 58], as well as on the links between generations of Aβ and tau hyperphosphorylations [16, 17, 59, 60]. According to our unifying hypothesis – built on the pivotal role of tau as a final common effector pathway – abnormal signaling leading to degenerative processes starts with the continuous activity of individually variable factors such as oxidative agents [61, 62], iron overload [63], disorders of lipid metabolism, hyperglycemia, deregulation of insulin levels, chronic infections, head trauma, and others [23]. These factors are likely to activate endogenous damage/alarm signals such as oxidized LDLs, oxyradicals, Aβ oligomers that trigger anomalous cellular signaling cascades in microglial cells and astrocytes. In turn, activated glial cells will respond by releasing NF-κB and overproduce proinflammatory cytokines (TNFα, Il-6, IL-1β), thus leading to cerebral inflammation [22, 64, 65] and serious alterations in neuron–glia interaction patterns [66, 67]. Interestingly, overproduction of IL-6 activates the JAK/Stat system via IL-6 and/or NMDA receptors, which in turn activates MAP kinases, thus promoting the activity of transcription factor Egr-1 that increases p35 expression (Fig. 1). This later effect results in the activation of cdk5, protein kinase involved in neuronal development [68], with subsequent tau hyperphosphorylation. The activation of the kinase p38 also results in tau hyperphosphorylations [59, 64]. Cytokines IL-6 and TNFα also appear to activate specific neuronal receptors, inducing cell cycle activity without proliferation. Interestingly, a dual dose-dependent action of TNFα producing either local neuroprotection or neuronal degeneration has been evidenced [65, 69]. One of the critical kinases involved in neuronal development, cdk5, is deregulated in AD according to in vitro evidence, studies in animal models with tau pathology [70], and studies in the human brain [71]. Hence, deregulation of cdk5 results in tau hyperphosphorylation at residues Ser_{202} and Thr_{205}. Inhibitors of cdk5 control tau hyperphosphorylation, neuronal degeneration, and neuronal death (Fig. 2). Thus, the deregulation of the sensitive equilibrium between protein kinases and phosphatases appears to be critical in the degenerative phenotype of neurons [62], which is reviewed in detail in an analysis of the involvement of cdk5 and gsk3β; by Maccioni et al. [2].

Anomalous cascades mediated by tau hyperphosphorylations have been also reported in several pathological pathways in neurons, such as p75 activated by an excess of NGF (or modified NGFs) produced by Aβ-activated glial cells [67], NO effects [66, 72–74]), direct action of Aβ oligomers on neuronal synapses [26], toxicity of advanced glycation end-products (AGES) via RAGE receptors [73], etc. downstream of the degenerative cascade, where hyperphosphorylated tau appears to be a final common pathway for neuronal degeneration associated with AD.

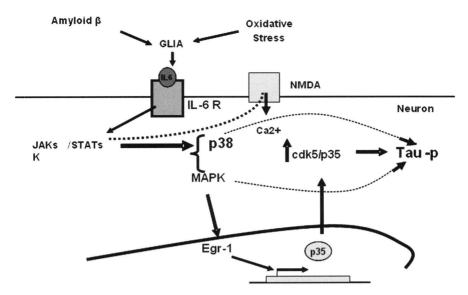

Fig. 1 Schematic representation of the effects of Aβ and brain oxidative stress signals (redox iron, oxygen, and nitrogen free radicals) on glial cells, the resulting release of IL-6 proinflammatory cytokine (could be also IL-1, TNFα), and activation of neuronal receptors and signaling through JAKs/STAT system. This signaling cascade activates MAPK and p38. Phosphorylated active MAPK, via the translation factor Erg-1, activates p35 gene, among others, increasing its neuronal expression, in the overactivation of cdk5 and the subsequent anomalous tau hyperphosphorylations (Representation generated from data of Quintanilla et al. [59] and Orellana et al. [64, 65]) (*See Color Plates*)

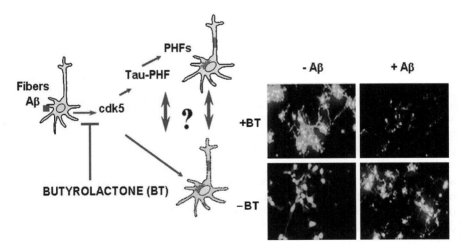

Fig. 2 Schematic representation showing the effects of inhibition of the protein kinase cdk5 by butyrolactone I (*BT*) on hippocampal cells in primary cultures. cdk5 inhibition protected hippocampal cells against the Aβ-induced neurodegeneration and neuronal death. The inmunofluorescence photomicrographs correspond to analysis of cell viability in the primary cultures of hippocampal cells derived from rat 18-day embryos. Studies indicate that cdk5 inhibitors such as BT (and roscovitine) protect neuronal cells against the neurotoxic effects of 10 μM soluble Aβ oligomers (Representation generated on the basis of experimental data from Alvarez et al. [16, 17] and Muñoz et al. [68]) (*See Color Plates*)

In summary, we propose that the long-term activation of the innate immune system by an individual (i.e., patient-specific) array of risk factors is a unifying mechanism triggering an altered signaling and inflammatory cascade that converges in tau hyperphosphorylation by protein kinase systems, cytoskeletal alterations (tau protein aggregation and PHFs formation), as our previously hypothesized final common pathway to latter lesions such as NFTs, interstitial amyloid protein deposits, and SP. In our view, the key pathogenic phenomena consist in the long-term of maladaptive activation of one or more innate immunity-triggering receptors – such as the toll-like receptors and the advanced glycation end-products receptors, and possibly also other unknown receptors located primarily in the microglial membrane – by a seemingly heterogeneous set of risk factors. In contrast with all previous hypotheses on AD, this view provides a unifying mechanism that explains both the diversity of the risk factors acting over long periods and the individual response to such insults. This formulation is susceptible of both empirical testing and its future implementation into novel therapeutic strategies of AD, as well as to other neurodegenerative disorders, in which impaired regulation of the innate immune system is the unifying cause of the condition.

5   Innovative Diagnostic Tools Based on Tau Pathology

Novel approaches to visualize and measure in vivo what is occurring in the brain, beyond the actual possibility to analyze *postmortem* brains [72, 75, 76, 77], are considered indispensable to attain significant progress in the diagnosis, prevention, and treatment of AD [78]. In fact, early detection of AD is widely felt to be critical in order to improve the efficacy in the design of future therapeutic approaches and to improve the quality of life of patients, considering that so far there is not appropriate molecular markers for neuroimaging. However, encouraging inroads have been made toward this goal. For example, a major recent finding was that there is a correlation between the increase in anomalously phosphorylated tau in the cerebrospinal fluid (CSF) and the degree of cognitive impairment in a subpopulation of patients with mild cognitive impairment and AD [28, 29]. Furthermore, a correlation with the ApoE4 allele was also found [27]. This type of correlations may one day serve as the basis for ongoing efforts to develop clinically useful in vivo imaging methodology for the cytoskeletal alterations in AD, and to correlate the latter with other parameters in the blood and CSF to make possible an early and perhaps even presymptomatic diagnosis of AD [29, 57, 78].

One of the most crucial yet unresolved aspects is the development of *pathology-specific neuroimaging technology* for the early detection of NFTs in the human brain in vivo [29, 79, 80]. The elucidation of tau structure and its conformational changes as a result of the interactions with specific ligands is critical for the development of PET technology to visualize tau aggregates in vivo, as a putative pathognomonic marker for the early diagnosis of AD [81]. Some ligands such as thioflavin (ThS) derivatives bind to isolated tau and PHFs, with a higher affinity than to the straight filaments (SF) found in AD [80]. The anticancer drug estramustine-P interacts with tau as well as with MAP-2 [9], and some benzimidazoles and benzo-thiazoles have affinity for the tau protein both in vitro and in vivo [82]. It has been shown that some benzimidazole derivatives tag the aggregated tau variants [57, 82, 83]. However, no structural studies have been carried out on tau interactions with these compounds, and no information exists on the specificity of the binding of the above-mentioned compounds with different conformational stages of tau proteins. Technological improvements will also contribute to the early diagnosis, and will be of help in monitoring patients during the development of novel disease-modifying therapies. This will facilitate the implementation of new therapeutic approaches, especially in cases in which neurodegenerative disorders can be corrected in an early phase, increasing the quality of life for patients and preventing the expenditure of resources for extended care of irreversible and debilitating disorders.

Sound advances have been attained in the last few decades in neuroimaging technology toward the diagnosis of neurological disorders. These have relied on the development of powerful computing applications and sophisticated software for image processing, along with increasing knowledge on the biology of neurological diseases. These neuroimaging techniques provide precise information on structural and functional aspects of the brain, but do not provide information on the type of specific pathology and its topographic distribution that is felt to be pathognomonic of individual diseases. In the field of neurodegenerative disorders, new neuroimaging methods have been developed that could be the basis for the design of a pathology-specific technology. To realize this promise in AD, it may be necessary to obtain maps of the lesions that are the major hallmarks of this disease, that is, the SP and the NFTs formed by hyperphosphorylated tau. Since Klunk and colleagues' publication [79] reporting the use of Pittsburgh compound-B as a radiotracer for the amyloid deposits in the human brain, a new era was launched in the development of in vivo neuroimaging of a pathology-specific process. At about the same time, Verhoeff et al. [84] published a similar study with a competitive radiotracer that may label both SP and NFTs. Considering that PHFs rather than SP are pathognomonics for AD, it is critical to investigate how to selectively tag NFTs in the brain of AD patients to visualize these lesions in vivo [57], and to fully realize the promise of this burgeoning technology.

6 Conclusion

In summary, the importance of tau alterations in AD became evident since Alois Alzheimer's discovery of the disorder in 1907 that now bears his name [1] –well before we knew of the existence of this protein – and remains central to the ongoing efforts to both understand and eventually conquer this devastating disorder. It is high time for this central role to be properly and widely acknowledged, as well as exploited toward the development of improved technology for the diagnosis, prevention, and treatment of AD and virtually all other neurodegenerative disorders characterized by alterations in the neuronal cytoskeleton. *The tau hypothesis of AD is and will remain pivotal in this fundamental effort*, and its implications are likely to loom large in the broad field of proteomics applied to human and veterinary diseases. The tau postulates together with the danger signal hypothesis on AD, indicating that innate immunity and inflammatory processes play a major role in triggering tau hyper-phosphorylations and the neurodegenerative cascade, conform an experimentally testable unifying hypothesis that accounts for the observations on the pathogenesis of AD. This is of high relevance toward the development of modern and reliable diagnostic tools and therapeutic approaches to control this devastating disease.

Acknowledgments Research was supported by grants from FONDECYT 1080254 and the International Center for Biomedicine and the Alzheimer's Association, U.S.A. to RBM. We appreciate the sound contributions of Dr. Jorge Fernandez, Leonardo Navarrete, and Karen Neumann.

References

1. Alzheimer A (1907) A singular disorder that affects the cerebral cortex. In: Hochberg CN, Hochberg FH (eds) Neurologic Classics in Modern Translation. Hafner Press, New York: Hafner Press, 1977, pp. 41–43
2. Maccioni RB, Concha I, Otths C, Muñoz JP (2001) The protein kinase cdk5: structural aspects, roles in neurogenesis and involvement in Alzheimer's pathology. Eur J Biochem 268:1518–1529
3. Maccioni RB, Barbeito L, Muñoz JP (2001) The molecular bases of Alzheimer's disease and other neurodegenerative disorders. Arch Med Res 32:367–381
4. Grundke-Iqbal I, Iqbal K, Tung YC, Quinlan M, Wisniewski HM, Binder LI (1986) Abnormal phosphorylation of the microtubule-associated protein tau (tau) in Alzheimer cytoskeletal pathology. Proc Natl Acad Sci USA 83:4913–4917
5. Kosik KS, Joachim CL, Selkoe DJ (1986) Microtubule-associated protein tau (tau) is a major antigenic component of paired helical filaments in Alzheimer disease. Proc Natl Acad Sci USA 83:4044–4048
6. Maccioni RB, Cambiazo V (1995) Role of microtubule-associated proteins in the control of microtubule assembly. Physiol Rev 75:835–864
7. Maccioni RB, Rivas CI, Vera JC (1988) Differential interaction of synthetic peptides from the carboxyl-terminal regulatory domain of tubulin with microtubule-associated proteins. EMBO J 7:1957–1963
8. Maccioni RB, Vera JC, Dominguez J, Avila J (1989) A discrete repeated sequence defines a tubulin binding domain on microtubule-associated protein tau. Arch Biochem Biophys 275:568–579
9. Moraga D, Rivas-Berrios A, Farias G, Wallin M, Maccioni RB (1992) Estramustine-phosphate binds to a tubulin binding domain on microtubule-associated proteins MAP-2 and tau. Biochim Biophys Acta 1121:97–103
10. Moraga DM, Nunez P, Garrido J, Maccioni RB (1993) A tau fragment containing a repetitive sequence induces bundling of actin filaments. J Neurochem 61:979–986
11. Cross D, Muñoz JP, Hernández P, Maccioni RB (2000) Nuclear and cytoplasmic tau proteins from human non-neuronal cells share common structural/functional features with brain tau. J Cell Biochem 78:305–317
12. González-Billault C, Farías G, Maccioni RB (1998) Modification of tau to an Alzheimer's type protein interferes with its interaction with microtubules. Cell Mol Biol 44:1117–1127

13. Farias GA, Vial C, Maccioni RB (1993) Functional domains on chemically modified tau protein. Cell Mol Neurobiol 13:173–182
14. Farias G, Gonzalez-Billault C, Maccioni RB (1997) Immunological characterization of epitopes on tau of Alzheimer's type and chemically modified tau. Mol Cell Biochem 168:59–66
15. Farias G, Muñoz JP, Garrido J, Maccioni RB (2002) Tubulin, actin and tau protein interactions and the study of their macromolecular assemblies. J Cell Biochem 85:315–324
16. Alvarez A, Toro R, Cáceres A, Maccioni RB (1999) Inhibition of tau phosphorylating kinase cdk-5 by butyrolactone and tau antisense probes prevents amyloid-induced neuronal death. FEBS Lett 459:421–426
17. Alvarez A, Muñoz JP, Maccioni RB (2001) A cdk5/p35 stable complex is involved in the beta-amyloid induced deregulation of cdk5 activity in hippocampal neurons. Exp Cell Res 264:266–275
18. Tabaton M, Mandybur TI, Perry G, Onorato M, Autilio-Gambetti L, Gambetti P (1989) The widespread alteration of neurites in Alzheimer's disease may be unrelated to amyloid deposition. Ann Neurol 26:771–778
19. Ghoshal N, García-Sierra F, Wuu J, Leurgans S, Bennett DA, Berry RW, Binder LI (2002) Tau conformational changes correspond to impairments of episodic memory in mild cognitive impairment and Alzheimer's disease. Exp Neurol 177:475–493
20. Sjoberg MK, Shestakova E, Mansoroglu Z, Maccioni RB, Bonnefoy E (2006) Tau protein binds to pericentromeric DNA: a putative role for nuclear tau in nucleolar organization. J Cell Sci. 119(Pt 10):2025–2034
21. Zhu X, Lee HG, Perry G, Smith MA (2007) Alzheimer disease, the two-hit hypothesis: an update. Biochim Biophys Acta 1772:494–502
22. Rojo L, Fernandez J, Maccioni AA, Jimenez J, Maccioni RB (2008) Neuro-inflammation: implications for the pathogenesis and molecular diagnosis of Alzheimer's disease. Arch Med Res 39:1–16
23. Fernandez J, Rojo LE, Kuljis RO, Maccioni RB (2008) The damage signal hypothesis of AD pathogenesis. J Alzheimers Dis 14:329–333
24. Glenner GG, Wong CW (1984) Alzheimer's disease and Down's syndrome: sharing of a unique cerebrovascular amyloid fibril protein. Biochem Biophys Res Commun 122:1131–1135
25. Hardy J, Selkoe DJ (2002) The amyloid hypothesis of Alzheimer's disease: progress and problems on the road to therapeutics. Science 297:353–356
26. Ferreira ST, Vieira MN, De Felice FG (2007) Soluble protein oligomers as emerging toxins in Alzheimer's and other amyloid diseases. Life 59:332–345
27. Lavados M, Farias G, Rothhammer F, Guillón M, Maccioni CB, Maccioni RB (2005) ApoE alleles and tau markers in patients with different levels of cognitive impairment. Arch Med Res 36:474–479
28. Maccioni RB, Lavados M, Maccioni CB, Mendoza A (2004) Biological markers of Alzheimer's disease and mild cognitive impairment. Curr Alz Res 1:307–314
29. Maccioni RB, Lavados M, Maccioni CB, Farias G, Fuentes P (2006) Anomalously phosphorylated tau protein and Abeta fragments in the CSF of Alzheimer's and MCI subjects. Neurobiol Aging 27:237–244
30. Damasio AR (1994) Descartes Error: emotion, Reason and the Human Brain. Avon Books
31. Rubia (2001) "El cerebro nos engaña?" Eds.: Ediciones temas de hoy, Madrid, pp. 339
32. Maccioni RB, Muñoz JP, Maccioni CB (2003) Dimensiones bioéticas de la investigación sobre el genoma humano (Bioethical dimensions of research on the human genome). Acta Bioética 10:75–81
33. Katzman R (1993) Education and the prevalence of dementia and Alzheimer's disease. Neurology 43:13–20
34. Khachaturian ZS (1998) An overview of Alzheimer's disease research. Am J Med 104(4A):26S–31S
35. Terry RD (2000) Where in the brain does Alzheimer's disease begin? Ann Neurol **47**:421

36. Hebert LE, Scherr PA, Bienias JL, Bennett DA, Evans DA (2003) Alzheimer disease in the US population: prevalence estimates using the 2000 Census. Arch Neurol 60:1119–1122
37. Delacourte A (2006) The natural and molecular history of Alzheimer's disease. J Alzheimers Dis 9(3 Suppl):187–194
38. Andreadis A (2005) Tau gene alternative splicing: expression patterns, regulation and modulation of function in normal brain and neurodegenerative diseases. Biochim Biophys Acta 1739:91–103
39. Kuljis RO (1994) Lesions in the pulvinar in patients with AD. J Neuropathol Exp Neurol 53:202–211
40. Kuljis RO, Tikoo RK (1997) Discontinuous distribution of senile plaques in striate hypercolumns in Alzheimer's disease. Vision Res 37:3537–3591
41. Duff K, Planel E (2005) Untangling memory deficit. Nat Med 11:826–827
42. Berger Z, Roder H, Hanna A et al. (2007) Accumulation of pathological tau species and memory loss in a conditional model of tauopathy. J Neurosci 27:3650–3662
43. Drewes G, Lichtenberg-Kraag B, Doring F et al. (1992) Mitogen activated protein (MAP) kinase transforms tau protein into an Alzheimer-like state. EMBO J 11:2131
44. Wille H, Drewes G, Biernat J, Mandelkow EM, Mandelkow E (1992) Alzheimer-like paired helical filaments and antiparallel dimers formed from microtubule-associated protein tau in vitro. J Cell Biol 118:573–584
45. Perez M, Arrasate M, Montejo E, de Garcini. Munoz V, Avila J (2001) In vitro assembly of tau protein: mapping the regions involved in filament formation. Biochemistry 40:5983–5991
46. Rivas CI, Vera JC, Maccioni RB (1988) Anti-idiotypic antibodies that react with microtubule-associated proteins are present in the sera of rabbits immunized with synthetic peptides from tubulin's regulatory domain. Proc Natl Acad Sci USA 85:6092–6096
47. Vera JC, Rivas CI, Maccioni RB (1988) Antibodies to synthetic peptides from the tubulin regulatory domain interact with tubulin and microtubules. Proc Natl Acad Sci USA 85:6763–6767
48. Cann JR, York EJ, Stewart JM, Vera JC, Maccioni RB (1988) Small zone gel chromatography of interacting systems: theoretical and experimental evaluation of elution profiles for kinetically controlled macromolecule-ligand reactions. Anal Biochem 175:462–473
49. Cross D, Vial C, Maccioni RB (1993) A tau like protein interacts with stress fibers and microtubules in human and rodent cultured cell lines. J Cell Sci 105:51–60
50. Henriquez JP, Cross D, Vial C, Maccioni RB (1995) Subpopulations of tau interact with microtubules and actin filaments in various cell types. Cell Biochem Funct 13:239–250
51. Cross D, Tapia L, Garrido J, Maccioni RB (1996) Tau-like proteins associated with centrosomes in cultured cells. Exp Cell Res 229:378–387
52. Sibille N, Sillen A, Leroy A et al. (2006) Structural impact of heparin binding to full-length tau as studied by NMR spectroscopy. Biochemistry 45:12560–12572
53. Appelt DM, Balin BJ (1993) Analysis of paired helical filaments (PHFs) found in Alzheimer's disease using freeze-drying/rotary shadowing. J Struct Biol 111:85–95
54. von Bergen M, Friedhoff P, Biernat J, Heberle J, Mandelkow EM, Mandelkow E (2000) Assembly of tau protein into Alzheimer paired helical filaments depends on a local sequence motif ((306)VQIVYK(311)) forming beta structure. Proc Natl Acad Sci USA 97:5129–5134
55. Mukrasch MD, Markwick P, Biernat J et al. (2007) Highly popsulated turn conformations in natively unfolded tau protein identified from residual dipolar couplings and molecular simulation. J Am Chem Soc 129:5235–5243
56. Maeda S, Sahara N, Saito Y, Murayama S, Ikai A, Takashima A (2006) Increased levels of granular tau oligomers: an early sign of brain aging and Alzheimer's disease. Neurosci Res 54:197–201
57. Rojo L, Avila M, Chandía M, Becerra R, Maccioni RB (2007) 18F Lanzoprazole: chemical and biological studies towards the development of new PET radiopharmaceuticals, International Conference on Clinical PET and Molecular Nuclear Medicine, 10–14 November 2007, Bangkok, Thailand
58. Rojo L, Sjoberg M, Hernandez P, Zambrano C, Maccioni RB (2006) Roles of cholesterol and lipids in the etiopathogenesis of Alzheimer's disease. J Biomed Biotech 2006:1–17

59. Quintanilla RA, Orellana DI, Gonzalez-Billault C, Maccioni RB (2004) Interleukin-6 induces Alzheimer-type phosphorylation of tau protein by deregulating the cdk5/p35 pathway. Exp Cell Res 295:245–257

60. Mendoza-Naranjo A, Gonzalez-Billault C, Maccioni RB (2007) Abeta1–42 stimulates actin polymerization in hippocampal neurons through Rac1 and Cdc42 Rho GTPases. J Cell Sci 120(Pt 2):279–288

61. Perry G, Castellani RJ, Hirai K, Smith M (1998) Reactive oxygen species mediate cellular damage in Alzheimer's disease. J Alzheimers Dis 1:45–55

62. Zambrano CA, Egana JT, Nunez MT, Maccioni RB, Gonzalez-Billault C (2004) Oxidative stress promotes tau dephosphorylation in neuronal cells: the roles of cdk5 and PP1. Free Radic Biol Med 36:1393–1402

63. Lavados M, Guillon M, Mujica MC, Rojo L, Fuentes P, Maccioni RB (2008) Mild cognitive impairment and Alzheimer's patients display different levels of redox-active CSF iron. J Alzheimers Dis 13:225–232

64. Orellana D, Quntanilla RA, Gonzalez C, Maccioni RB (2005) Role of JAKs/STAT pathway in the intracellular calcium changes induced by interleukin-6 in hippocampal neurons. Neurotoxicity Res 8:295–304

65. Orellana DI, Quintanilla RA, Maccioni RB (2007) Neuroprotective effect of TNFalpha against the beta-amyloid neurotoxicity mediated by cdk5 kinase. Biochim Biophys Acta 1773:254–263

66. Saez M, Pehar M, Barbeito L, Maccioni RB (2004) Astrocytic nitric oxide triggers tau hyperphosphorylation in hippocampal neurons. In Vivo 18:275–280

67. Saez M, Pehar M, Vargas M, Barbeito L, Maccioni RB (2006) Production of NGF by β-amyloid stimulated astrocytes induces p75NTR-dependent tau hyperphosphorylation in cultured hypocampal cells. J Neurosci Res 84:1098–1106

68. Muñoz JP, Alvarez A, Maccioni RB (2000) Regulation of the expression of the cyclin-dependent protein kinase cdk5 during laminin-induced neuritic development in neuroblastoma cells. Neuroreport 11:12–21

69. Tarkowski E, Blennow K, Wallin A, Tarkowski A (1999) Intracerebral production of TNF-alpha, a local neuroprotective agent, in Alzheimer disease and vascular dementia. J Clin Immunol 19:223–230

70. Otth C, Concha II, Arendt T et al. (2002) AbetaPP induces cdk5-dependent tau hyperphosphorylation in transgenic mice Tg2576. J Alzheimers Dis 4:417–430

71. Patrick GN, Zukerberg L, Nikolic M, de la Monte S, Dikkes P, Tsai LH (1999) Conversion of p35 to p25 deregulates cdk5 activity and promotes neurodegeneration. Nature 402:615–622

72. Fischer HC, Kuljis RO (1994) Multiple type of nitrogen monoxide synthase/NAPH diaphorase containing neurons in the human cerebral neocortex. Brain Res 654:105–107

73. Hernanz A, De la Fuente M, Navarro M, Frank A (2007) Plasma aminothiol compounds, but not serum tumor necrosis factor receptor II and soluble receptor for advanced glycation end products, are related to the cognitive impairment in Alzheimer's disease and mild cognitive impairment patients. Neuroimmunomodulation 14:163–167

74. Kuljis RO, Shapshak P, Alcabes P, Rodríguez de la Vega P, Fujimura R, Petito CK (2002) Increased density of neurons containing NADPH diaphorase and nitric oxide synthase in the cerebral cortex of patients with HIV-1 infection and drug abuse. J NeuroAIDS 2:19–36

75. Kuljis RO (1997) Modular corticocerebral pathology in Alzheimer's disease. In: Mangone CA, Allegri RF, Arizaga RL, Ollari JA (eds) Dementia: a Multidisciplinary Approach. Editorial Sagitario, Buenos Aires, Argentina, pp. 143–155.

76. Kuljis RO, Schelper RL (1996) Alterations in nitrogen monoxide-synthesizing cortical neurons in amyotrophic lateral sclerosis with dementia. J Neuropathol Exp Neurol 55:25–35

77. Kuljis RO (2008) Alzheimer's disease (updated entry). eMedicine (available online at www. eMedicine.com)

78. Teunissen CE, de Vente J, Steinbusch HW, De Bruijn C (2002) Biochemical markers related to Alzheimer's dementia in serum and cerebrospinal fluid. Neurobiol Aging 23:485–508

79. Klunk WE, Engler H, Nordberg A, et al. (2004) Imaging brain amyloid in Alzheimer's disease with Pittsburgh Compound-B. Ann Neurol 55:306–319
80. Santa-Maria I, Perez M, Hernandez F, Avila J, Moreno FJ (2006) Characteristics of the binding of thioflavin S to tau paired helical filaments. J Alzheimers Dis 9:279–285
81. von Bergen M, Berghorns S, Biernat J, Mandelkow EM, Madelkow E (2005) Tau aggregation is driven from a transition coil to beta sheet structure. Biochem Biophys Acta 3:1739
82. Okamura N, Suemoto T, Furumoto S et al. (2005) Quinoline and benzimidazole derivatives: candidate probes for in vivo imaging of tau pathology in Alzheimer's disease. J Neurosci 25:10857–10862
83. Mathis CA, Klunk WE, Price JC, DeKosky ST (2005) Imaging technology for neurodegenerative diseases: progress toward detection of specific pathologies. Arch Neurol 62:196–200
84. Verhoeff NP, Wilson AA, Takeshita S, et al. (2004) In-vivo imaging of Alzheimer disease beta-amyloid with [11C] SB-13 PET. Am J Geriatr Psychiatry 12:584–595

Neuronal Cytoskeleton Regulation and Neurodegeneration

Ya-Li Zheng, Niranjana D. Amin, Parvathi Rudrabhatla, Sashi Kesavapany, and Harish C. Pant

Abstract The biology of neurodegeneration program evolved from the laboratory studying the basic biology of neuronal cytoskeletal protein phosphorylation during development and normal function in the adult. To understand the molecular basis of neurodegeneration, our major focus has been to study the regulation of compartment-specific patterns of cytoskeletal protein phosphorylation in neuronal perikarya and axons. We have demonstrated that the phosphorylation of the numerous acceptor sites on proline-directed serine and thronine (Pro-Ser/Thr) residue proteins such as tau and neurofilaments is tightly regulated. The phosphorylation of these molecules is generally confined to the axonal compartment. It was recognized that in neurodegenerative disorders such as Alzheimer's disease (AD) and amyotrophic lateral sclerosis (ALS), the pathology was characterized by an accumulation of aberrantly phosphorylated cytoskeletal proteins in cell bodies, suggesting that topographic regulation had been compromised. This led inevitably into studies of neurodegeneration in cell culture and model mice with emphasis on a specific neuronal protein kinases, for example, cyclin-dependent kinase 5 (Cdk5), that target numerous neuronal proteins including cytoskeletal proteins, which when deregulated may be responsible for the pathology seen in neurodegeneration. In cell systems, neuronal stress leads to deregulated kinases, for example, Cdk5, accompanied by abnormal cytoskeletal protein phosphorylation and cell death characteristic of neurodegeneration. In this chapter, efforts are made to answer some of the following questions. (1) How is the cytoskeletal protein phosphorylation topographically and stably regulated in their proline-directed Ser/Thr residues in neurons? (2) What factors are responsible for the regulation and deregulation of Cdk5 in neurons?

Y.-L. Zheng, N.D. Amin, P. Rudrabhatla, S. Kesavapany, and H.C. Pant (✉)
Cytoskeletal Protein Regulation Section
Laboratory of Neurochemistry
NINDS, NIH,
Bethesda, MD 20892
e-mail: panth@ninds.nih.gov

R.B. Maccioni and G. Perry (eds.) *Current Hypotheses and Research Milestones in Alzheimer's Disease*. DOI: 10.1007/978-0-387-87995-6_6,
© Springer Science + Business Media, LLC 2009

Abbreviations Cdk5: cyclin-dependent kinase 5; EGF: epidermal growth factor; GSK3: glycogen synthase kinase 3; Erk1/2: extracellular signal-regulated kinases 1 and 2; MAPK: mitogen-activated protein kinase; Pin1: protein interacting with NIMA (never in mitosis A)-1; p-NF-H: phosphorylated neurofilament-high molecular weight subunit.

1   Background and Significance

In a mature nervous system, neuronal cytoskeleton phosphorylation is topographically regulated [1], and phosphorylation of cytoskeletal proteins is restricted to axonal but not cell body compartment under normal conditions. Figure 1a represents a cartoon view of topographic regulation of neuronal cytoskeletal organization. Although kinases, phosphatases, cytoskeletal protein substrates, and their regulators are synthesized in cell bodies, the phosphorylation of cytoskeletal proteins, for example, tau, medium molecular mass (neurofilament-medium molecular weight subunit, NF-M), and high molecular mass (NF-high molecular weight subunit, NF-H) tail domains, occurs selectively and stably in the axonal compartment during axon transport [2, 3]. Figure 1b illustrates the selective phosphorylation of cytoskeletal proteins in neurites in human cervical spinal cord neurons under physiological conditions. In several neurodegenerative disorders, however, such as Alzheimer's disease (AD) and amyotrophic lateral sclerosis (ALS), this regulation is deregulated, and an aberrant and stable phosphorylation of cytoskeletal proteins is found in the cell body compartment (Fig. 1c). The mechanisms of topographic regulation and deregulation are not well understood. For several years, our major focus has been the study of the kinases that regulate the phosphorylation of NFs and other proteins in neurons [4].

NFs are the major cytoskeletal component of large myelinated axons in the mammalian nervous system and phosphorylated in the axonal compartment. Together with microtubules, mitogen-activated proteins (MAPs) and associated molecules, NFs, contribute to axonal morphology and function. Cytoskeletal protein function is markedly affected by its state of phosphorylation, which is regulated by reciprocally interacting kinases and phosphatases. The normal phosphorylation of NFs provides stability to axonal structure, protects from proteolysis, regulates axonal transport, and determines axonal caliber, thereby regulating conduction velocity. To identify the active players, we studied the kinases responsible for the multisite phosphorylation of lysine/serine/proline (KSP) repeats in the C-terminal tail domains of NF-H and NF-M neurofilament proteins and tau [4]. It was found that mitogen-activated protein kinase (MAPK) (Erk1/2; SAPK) and cyclin-dependent kinase 5 (Cdk5) were the major kinases responsible for KSP multiple site phosphorylation. We purified these kinases from nervous tissues, and demonstrated their phosphorylation of KSP repeat motifs in C-terminal tail domains of NF-M/H. Cdk5, the major kinase, was cloned from a rat brain cDNA library and its primary structure and similarities with other Cdks was studied [5, 6]. It soon became clear that the proline-directed kinases

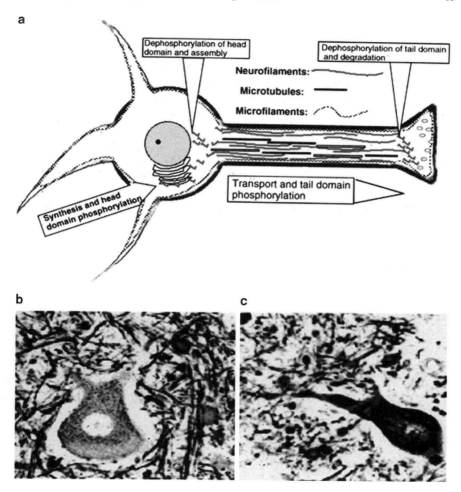

Fig. 1 a Represents a diagrammatic view of topographic phosphorylation of neuronal cytoskeletal proteins, biosynthesis in the cell body, assembly of cytoskeletal proteins in axon-hillock region, phosphorylation and transport in the axonal compartment, and finally dephosphorylation and degredation at the nerve terminals. **b** Human cervical spinal cord neuron (physiology), no phosphorylation occurs in the cell body; however, neurites are selectively phosphorylated (SMI 31 staining). **c** Human cervical spinal cord neuron from ALS patient (pathology). Aggregates of aberrantly hyperphosphorylated deposit cytoskeletal proteins in the cell body (*See Color Plates*)

(Cdk5 and Erk1/2), together with other kinases such as casein kinase I,2 (CKI, CK2), glycogen synthase kinase 3 (GSK3), protein kinase A (PKA), played a key role in the phosphorylation of most cytoskeletal proteins including microtubules, tau, and MAPs [4]. The specificity of Cdk5 and Erk1/2 phosphorylation of KSP repeats in NFs depended on its motif structure, which varied from species to species. For example,

Erk1/2 was the major kinase phosphorylating rat and mouse NF-M/NF-H, while human NF-H and tau were the major target for Cdk5 [7].

To understand the deregulation of cytoskeleton phosphorylation inducing pathology, it is important to know how the cytoskeletal protein phosphorylation is regulated under physiological conditions in the neurons, for example, what are the mechanisms of topographic phosphorylation in the neurons? To answer these questions, we proposed the following hypotheses:

Hypothesis (1)

Initially, the unphosphorylated tail domains of NF-M/H are compact globular appendage to the C-terminus of the rod domain in NF-M/H components and not exposed to the proline-directed Ser/Thr kinases. We proposed that the transient head domain phosphorylation by PKA, PKC, or CaMPKs of NF-L-M/H in the cell body inhibits the tail domain KSP phosphorylation.

Hypothesis (2)

The phosphorylation of neuronal cytoskeletal proteins is differentially regulated in the cell body and axonal compartments; higher in axons compared to cell body.

Hypothesis (3)

After entry of dephosphorylated NF/MT-complex into the axon hillock, a few Ser/Thr-Pro (S/T-P) sites become accessible to proline-directed Ser/Thr kinases, for example, MAPKs and Cdk5 which are activated by glial/axonal interactions and phosphorylate the exposed few S/T-P motifs of cytoskeletlal proteins during axonal transport.

Hypothesis (4)

Protein interacting with NIMA (never in mitosis A)-1 (Pin1), a peptidyl-prolyl *cis/trans* isomerase, induces stabilization of the phosphorylated S/T-P motifs in the proteins.

Hypothesis (5)

Hyperactivation of proline-directed Ser/Thr kinases and their involvement in neurodegenerative diseases.

1.1 Hypothesis (1)

Figure 2a shows the NF-subunits and their phosphorylation domains phosphorylated by identified kinases. To test this hypothesis, we used NF-M. Figure 2b represents NF-M head domain residues phosphorylated by PKA and tail domain by CK1/2 and MAPKs (Erk1/2). Bacterially expressed and purified NF-M was phosphorylated in the presence of PKA (1), Erk1/2 (2), or phosphorylated first by PKA then Erk1/2 (3). PKA phosphorylates head domain while Erk1/2 phosphorylates KSP

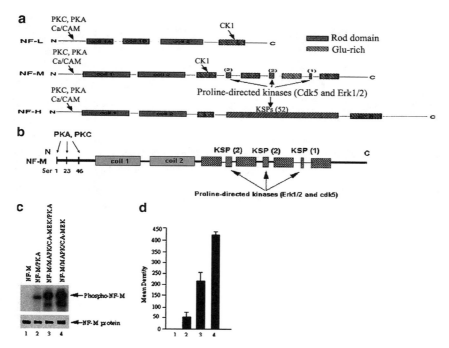

Fig. 2 **a** Domain structures of NF-subunit proteins, NF-L (low molecular weight), NF-M (middle molecular weight), NF-H (high molecular weight), their phosphorylation domains by identified kinases, PKA (cAMP-dependent protein kinase), PKC (protein kinase C), Ca^{2+}/CAM (calcium–calmodulin-dependent kinase), proline-directed Ser/Thr kinases (Cdk5, Erk1/2, GSK3β), casein kinase I (CK I), and casein kinase II (CK II). **b** Phosphorylated residues in the head domain of NF-M by PKA. (**c**) Phosphorylation of bacterially expressed and purified NF-M in vitro. *Lane 1:* No kinase added, only NF-M, *Lane 2:* NF-M phosphorylated by PKA, *Lane 3:* NF-M first phosphorylated by PKA, then Erk1/2, and *Lane 4:* NF-M phosphorylated by Erk1/2. *Top panel* is autoradiograph and *bottom* is Commassie stain. **d** represents the quantitation of data shown in **c** (*See Color Plates*)

repeats in the tail domain. As shown in Fig. 2c and d, the head domain phosphorylation of NF-M by PKA inhibited the tail domain phosphorylated by MAPKs (compare lanes 2 and 3). In addition, nonneuronal cells transfected with NF-M and stimulated by either with or without PKA activator, EGF, provided the similar results [8]. These studies suggest that the head domain phosphorylation inhibits the tail domain KSP phosphorylation in NF-M.

1.2 Hypothesis (2)

To test this hypothesis, we used squid giant axon system shown in Fig. 3a. Axonal components (squeezed axioplasm) free from the cell body components and cell body material free from axonal components can be isolated from this preparation

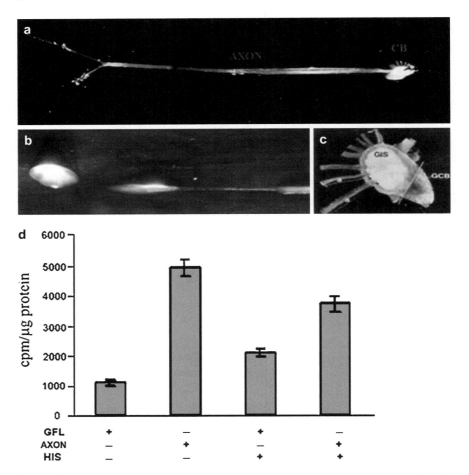

Fig. 3 a Squid giant axon containing ganglion cell body. **b** Squeezed out axioplasm. **c** Ganglion cell body bag. **d** Endogenous and exogenous phosphorylation activity of axon and cell body preparations, endogenous (no histone H1) and exogenous (presence of histone H1) (*See Color Plates*)

(Fig. 3b and c). The phosphorylation activity of proteins in these compartments is compared. The data presented in Fig. 3d demonstrate that the phosphorylation activity is higher in axonal than cell body compartment. In addition, the tyrosine phosphatase activity was higher in cell body compared to axon [1], but Ser/Thr phosphatase activity was found higher in the axonal compartment [1]. This is consistent with data that there is no stable head domain phosphorylation found in NF-M or other NF-subunits [9, 10]. The head domains of NF-subunits are transiently phosphorylated in the cell body, but not in the axonal compartment.

1.3 Hypothesis (3)

One candidate for glial/axonal interaction is myelin-associated glycoprotein (MAG), located in the periaxonal region, that may regulate the expression and phosphorylation of cytoskeletal molecules in the axonal compartment (Fig. 4a). The most convincing demonstration of the importance of localized signals in modulating the axonal phosphorylation of NF and MAPs comes from a study of the effect of MAG on phosphorylation of cytoskeletal proteins in DRG and PC12 cells cocultured with COS-7 cells transfected with MAG [11] (Fig. 4b–d). It has previously been demonstrated that myelination promotes NF phosphorylation and organization within axons [12, 13]. The intimate glial–axon interaction involves mutual signaling and one consequence of myelination is phosphorylation of axonal NF and MAPs. In these coculture situations, the kinases Cdk5 and Erk1/2 were upregulated, accompanied by an increased amount of NF-M, NF-H, tau, and MAP1B and MAP2

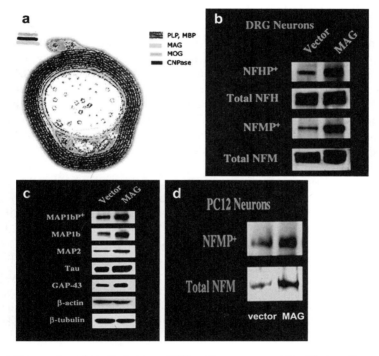

Fig. 4 **a** Cross section of myelinated axon. Different layers represent identified myelin proteins. **b** Dorsal root ganglia isolated from rat were cocultured with COS cell containing empty vector (*left lanes*) and MAG stably transfected (*right lanes*), expression of NF-M/H was analyzed using phospho- and total NF-M/H-specific antibodies. **c** Expression of MAP1b, MAP2, Tau, GAD-43, β-actin, and tubulin was analyzed using respective specific antibodies to corresponding molecules. **d** PC-12 cells were cocultured with COS cell and transfected with vector alone (*left lanes*) and MAG-containing vector (*right lanes*) expression of phosphorylated and total NF-M was analyzed (*See Color Plates*)

as well as increased phosphorylated NF-H, NF-M, and MAP1B. These data suggest that axonal NF and MAP phosphorylation is induced by glial–axonal interactions via MAG activation of Erk1/2 and Cdk5 kinase cascades [11]. These studies are confirmed using DRG cells from MAG knockout (KO). There is a drastic reduction in Cdk5 and MAPK (Erk1/2) activity and cytoskeletal protein phosphorylation in KO compared to wild type. Also, other exogenous signals (integrin-mediated cell adhesion, Ca^{2+} influx, trophic factors, such as EGF, retinoic acid, and glial-derived factors) were demonstrated to activate Cdk5 and/or Erk1/2 pathways and stimulated NF and MAP protein phosphorylation in transfected cells and in primary neurons in culture [11, 14–18]. These signal transduction cascades correlated with and seemed to be responsible for the phosphorylation and formation of a functional cytoskeleton during neurite outgrowth, axon elongation, and stabilization.

1.4   Hypothesis (4)

Within the last decade, a novel level of modulation of protein phosphorylation has emerged, namely, factors that regulate the conformation and stability of proteins phosphorylated at S/T-P sites by proline-directed kinases. Most proline-directed Ser/Thr kinases [19] and phosphatases [20] are highly selective for *trans* S/T-P bonds. Peptidyl-prolyl *cis/trans* isomerases such as Pin1 specifically target phosphorylated S/T-P sites and by virtue of the proline residue can "toggle" an inactive *cis*-isomer to the more stable *trans*-form, with altered function. Pin1 plays a key role in diverse cellular functions, including the cell cycle, cancer, neurodegeneration, and apoptosis [19–21]. Pin1 is localized in nuclei of most cells, where it modulates the functions of several mitotic proteins. In neurons, however, Pin1 is distributed in both nucleus and cytoplasm, increases during neuronal differentiation, and its expression correlates with the phosphorylation of tau at a specific Thr 231 site [21] and Pin1 is detected in neurofibrillary tangles in AD brains. Like tau, NFs contain many Ser/Thr phosphate acceptor sites that are targeted by several proline-directed kinases and phosphatases. In contrast to tau, however, NFs, particularly NF-H, are enriched with numerous KSP repeat motifs (43–100), depending on species, in the tail domain, sites for proline-directed kinase phosphorylation. The tail domain KSP repeats of rat NF-H is shown in Fig. 5a. Most of all the Ser/Thr residues in KSP motifs are phosphorylated in vivo [9, 10] (Fig. 5b shows a schematic representation of Pin1). WW domain of Pin1 binds with pS/T-P motifs and induces *cis/trans* isomerization.

The question arises, does Pin1 play any role in stabilizing the numerous phosphorylated KSP sites in NFs? Since hyperphosphorylated NF proteins are one of the principal components of perikaryal aggregates in ALS and AD, we studied ALS spinal cord and AD brain tissues for Western blot and immunocytochemical analyses and found that Pin1 indeed coprecipitates with phosphorylated NF-H (p-NF-H), and is expressed at higher levels than in control tissues, while soluble Pin1 remains constant [22] (Fig. 5c and d). This was true for both AD and ALS tissues suggesting

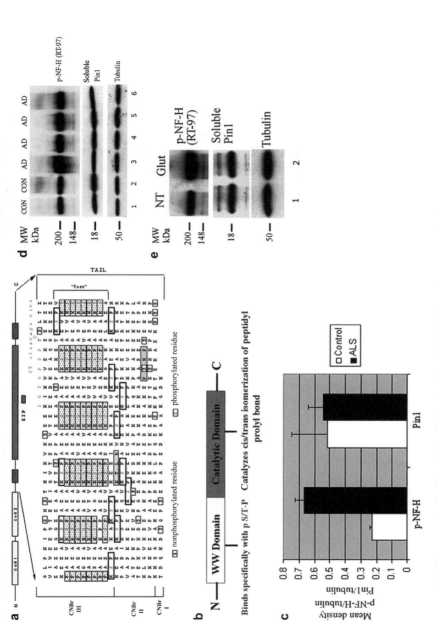

Fig. 5 a Represents the KSP repeats in the C-terminal tail domain of rat NF-H. **b** Schematic presentation Pin1 domains. **c** Expression of NF-H (phosphorylated) and Pin1 in ALS and control human brains. **d** Represents the expression of NF-H (phosphorylated), Pin1 (soluble), and tubulin from AD and control brains. **e** Glutamate stress results in the increase of p-NF-H in the neurons. *Lane 1*, control and *Lane 2*, glutamate-stressed neurons. p-NF-H (*upper panel*), soluble Pin1 (*middle panel*), and tubulin (*lower panel*) are shown

that Pin1 may, as it does for tau, also modulate NF phosphorylation in neurodegeneration. Moreover in immunocytochemical assays of ALS spinal cords, Pin1 colocalized with p-NF-H in aggregates in the ventral horn cells [22]. This suggests that Pin1 may play a role in regulating NF-H phosphorylation, particularly at the numerous KSP tail domain sites and may contribute to the neuronal pathology.

To explore this question, we resorted to GST-Pin1 pull down assays of rat brain lysates Coomassie-stained gels, and Western blots were prepared using an antibody specific for p-NF-H (RT-97). A Coomassie band presumed to be NF-H was excised and identified mass spectrometrically as p-NF-H. Pin1 and p-NF-H also coimmunoprecipitate from rat brain lysates, linking Pin1 to p-NF-H tail domain KSP repeats. These data suggest that Pin1 binds to p-NF-H. The question is, what role does it play and how? As an initial hypothesis, we suggest that Pin1 is essential in stabilizing the phosphorylated KSP repeats in the tail domain as they are being phosphorylated by proline-directed kinases during axonal transport.

Neuronal stress (e.g., oxidative, excitotoxic) deregulates (hyperactivities) the activities of kinases that tightly regulate topographic phosphorylation. This deregulation involves abnormal activation of proline-directed kinases such as Cdk5, Erk1/2, SAPK, and p38, and results in hyperphosphorylated and aberrant cytoskeletal proteins, for example, p-NF-H and tau, accumulation within perikarya, leading to cell death [23–25]. Since Pin1 is expressed in neuronal cell nuclei and cytoplasm, stress upregulation of proline-directed kinases within cell bodies may also evoke a Pin1 response and stimulate aberrant NF-H phosphorylation. To test this, we used primary rat dorsal root ganglion (DRG) neurons (E16–18) in culture for several days challenged with an excitotoxic stress stimulus (glutamate treatment) which is known to induce perikaryal phosphorylated-tau and NF accumulations and cell death [23]. Our studies show that excitotoxic stimulation produced activation of proline-directed kinases and elevated the levels of p-NFH significantly in the stressed neurons compared to controls (Fig. 5e). In addition, it was found that immunocytochemical assays showed that Pin1 and p-NFH are colocalized in perikarya of stressed neurons, resembling the pathology in ALS motor neurons [22]. Both effects were rescued by prior treatment of cells with juglone, the Pin1-specific inhibitor or dominant negative Pin1 [22]. Although Pin1 is expressed in cell bodies, these results would suggest that stress-induced elevation of proline-directed kinases in the cell body is primarily responsible for aberrant phosphorylation of KSP sites while Pin1 acts to stabilize them as they are formed.

1.5   Hypothesis (5)

Neurodegeneration associated with aberrant and deregulation of neuronal cytoskeletal protein phosphorylation is due to hyperactivation of proline-directed Ser/Thr kinases and their involvement in neurodegenerative diseases. Oxidative stress signals as well as amyloid-β (Aβ) oligomers and polymers do activate Cdk5, an event that determines specific tau hyperphosphorylation. One of the first evidence in this respect came in 1998 from studies on Cdk hyperactivation in hippocampal cells [26–28] and the signaling involved in tau hyperphosphorylation by other kinases such as p38 [29].

It has been found that there is an excessive phosphorylation of S/T-P residues of many proteins in several degenerative diseases. This suggests that the proline-directed Ser/Thr kinases are responsible for their phosphorylation. We have studied Cdk5, a neuron-specific proline-directed Ser/Thr kinase and found that the regulation and deregulation of Cdk5 is associated with neuronal survival and death.

Cdk5 is one of the major kinase phosphorylating the neuronal cytoskeletal proteins in the nervous system. It differs from other cell cycle kinases. Though a member of the family of Cdks, it is inactive in the cell cycle, but active primarily in postmitotic neurons in association with neuron-specific regulators, p35, p39 [30], and p67 [31]. By virtue of its phosphorylation of diverse substrates, Cdk5 is a multifunctional kinase. It is involved in neurogenesis, neuronal migration, synaptic activity, cell survival, and even enzymes of signal transduction pathways [32]. Its role in neural development was clearly demonstrated in Cdk5 and p35 KO mice. Although the former phenotype was lethal at birth and the latter survived to adulthood, both exhibited similar defects in cortical layering, cerebellar foliation, and axonal fasciculation [33, 34]. Subsequently, it was shown that expression of a Cdk5 using p35, a neuron-specific promoter in the Cdk5–/– background, rescued the Cdk5 KO phenotype. In addition, the double KO, p35–/– and p39–/– produced a phenotype identical to the Cdk5 KO. These studies suggest that Cdk5 activity is responsible for neuronal survival [35].

2 Role of Cdk5 in Neuronal Survival and Neurodegeneration

Meanwhile Cdk5 has been identified as a tightly regulated multifunctional kinase in the nervous system, important in neuronal migration during development and essential for survival (Fig. 6a). When Cdk5 is deregulated in neuronal stress and/or neural degeneration, it can lead to abnormal perikaryal accumulations of phosphorylated cytoskeletal proteins such as tau and NFs and to cell death. This links Cdk5 directly to the problem of topographic regulation of cytoskeletal protein phosphorylation, and raises questions as to its role in neuronal survival. A model of Cdk5 deregulation proposes the stress-induced cleavage of the p35 activator to the more stable and hyperactivating regulator p25, which promotes abnormal perikaryal tau and NF hyperphosphorylation (Fig. 6a). As indicated above, the tightly regulated, multifunctional kinase, Cdk5, plays a key role in regulating topographic phosphorylation of neuronal cytoskeletal proteins. When deregulated as above, it has been implicated as a key player leading to neurodegenerative pathologies in disorders such as AD, ALS, Parkinson's disease, and Nieman-Pick disease [36, 37]. Amyloid plaques and neurofibrillary tangles have been linked to deregulated Cdk5 and neuronal death. Aβ, being toxic, may evoke abnormal calcium influx which will activate calpains followed by cleavage of p35 to p25 which, in turn, mislocalizes and hyperactivates Cdk5 phosphorylation of tau and NFs in cell bodies.

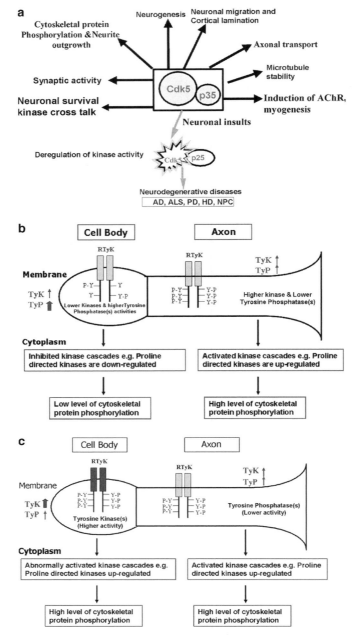

Fig. 6 a Cdk5 is a multifunctional protein kinase. Some of the neuronal processes regulated and studied in our laboratory are shown. Upon deregulation, p35 is cleaved by calpain and p25/Cdk5 is hyperactivated and induces neurodegeneration by aberrant and hyperphosphorylation of neuronal cytoskeletal and other proteins. **b** Hypothetical model; tightly regulated topographic phosphorylation processes in the neuron are controlled by tightly and topographically regulated kinases and phosphatases. **c** Aberrant and hyperactivity of proline-directed Ser/Thr kinases deregulate the topographic phosphorylation of cytoskeletal and other proteins and induce pathology

3 Hypothesis (6)

Protein phosphatases are essential to nervous system development and function. Phosphatases (dephosphorylating Ser/Thr and Tyr sites) are critical in the regulation of signal transduction cascades [38, 39] and modulate synaptic protein phosphorylation and function in neurons [40, 41]. Along with kinases and other proteins, they are organized into multimeric receptor complexes involved in synaptic transmission and LTP [42]. As protein tyrosine phosphatase receptors, they play a key role regulating cytoskeletal dynamics during axonal outgrowth [43]. The Ser/Thr phosphatase family, including PP1 and PP2, consists of a catalytic and regulatory subunit, the latter determining substrate specificity and intracellular localization [44]. Phosphatases are bound to NFs and tau in cytoskeletal preparations [45–47] and are localized within different cellular compartments by cytosolic or cytoskeletal-bound targeting proteins [47]. No doubt, they are active players in topographic regulation of neuronal phosphorylation and neurodegenerative diseases.

The squid giant fiber system is ideal for studies of compartment-specific patterns of phosphorylation because cell bodies and pure axoplasm are easily separated for biochemical studies. Although cell body (GFL) extracts contain active kinases, endogenous cytoskeletal protein phosphorylation is inhibited compared to the more active axoplasm [1]. This suggests the presence of kinase inhibitors or higher phosphatase levels. In preliminary studies, we have shown significant stimulation of GFL phosphorylation by vanadate, a tyrosine phosphatase inhibitor. Axoplasm extracts, on the other hand, were more responsive to okadaic acid, an Ser/Thr phosphatase inhibitor. Is the low endogenous phosphorylation activity of cell body extracts due to higher phosphatase activities? Are the phosphatases modulated in neurodegenerative disorders? To further analyze the role of phosphatases in topographic regulation of phosphorylation, it is essential to understand the topographic cytoskeleton phosphorylation and neurobiology of neurodegeneration. Squid giant axon may provide an ideal model system.

A hypothetical proposal shown in Fig. 6b and c illustrates that the topographic regulation of cytoskeletal protein phosphorylation is a tight regulation among kinase and phosphatase activities involved in neuronal survival and death. The deregulation of these processes results in aberrant and hyperphosphorylation of neuronal cytoskeletal proteins in neurodegeneration.

4 Conclusion

The aberrant and hyperphosphorylation of neuronal cytoskeletal proteins in neurodegenerative disorders such as AD, PD, and ALS is mainly due to the deregulation of proline-directed kinases and phosphatases. We proposed that the transient head domain phosphorylation of NF-L-M/H by PKA, PKC, or CaMPKs in the cell body inhibits the tail domain KSP phosphorylation. The phosphorylation of neuronal cytoskeletal proteins is differentially regulated in the cell body and axonal compartments; higher in axons compared to cell body. The tightly regulated, multifunctional kinase, Cdk5, plays a key role in regulating topographic phosphorylation of neuronal cytoskeletal proteins. Oxidative stress signals as well as Aβ oligomers and polymers do activate Cdk5, an event that determines specific NF and tau hyperphosphorylation. Finally, prolyl isomerases such as Pin1 modulates the stress-induced perikaryal hyperphosphorylation of KSP repeats of NFs.

Acknowledgments This work is supported by the Intramural Research Program of the NINDS, National Institutes of Health.

References

1. Grant P, Diggins M, Pant HC (1999) Topographic regulation of cytoskeletal protein phosphorylation by multimeric complexes in the squid giant fiber system. J Neurobiol 40:89–102
2. Nixon RA, Paskevich PA, Sihag RK, Thayer CY (1994) Phosphorylation on carboxyl terminus domains of neurofilament proteins in retinal ganglion cell neurons in vivo: influences on regional neurofilament accumulation, interneurofilament spacing, and axon caliber. J Cell Biol 126:1031–1046
3. Pant HC, Veeranna (1995) Neurofilament phosphorylation. Biochem Cell Biol 73:575–592
4. Grant P, Pant HC (2000) Neurofilament protein synthesis and phosphorylation. J Neurocytol 29:843–872
5. Hellmich MR, Pant HC, Wada E, Battey JF (1992) Neuronal cdc2-like kinase: a cdc2-related protein kinase with predominantly neuronal expression. Proc Natl Acad Sci USA 89:10867–10871
6. Shetty KT, Kaech S, Link WT et al. (1995) Molecular characterization of a neuronal-specific protein that stimulates the activity of Cdk5. J Neurochem 64:1988–1995
7. Pant AC, Veeranna, Pant HC, Amin N (1997) Phosphorylation of human high molecular weight neurofilament protein (hNF-H) by neuronal cyclin-dependent kinase 5 (cdk5). Brain Res 765:259–266
8. Zheng YL, Li BS, Veeranna, Pant HC (2003) Phosphorylation of the head domain of neurofilament protein (NF-M): a factor regulating topographic phosphorylation of NF-M tail domain KSP sites in neurons. J Biol Chem 278:24026–24032
9. Jaffe H, Veeranna, Pant HC (199a) Characterization of serine and threonine phosphorylation sites in beta-elimination/ethanethiol addition-modified proteins by electrospray tandem mass spectrometry and database searching. Biochemistry 37:16211–16224
10. Jaffe H, Veeranna, Shetty KT, Pant HC (199b) Characterization of the phosphorylation sites of human high molecular weight neurofilament protein by electrospray ionization tandem mass spectrometry and database searching. Biochemistry 37:3931–3940
11. Dashiell SM, Tanner SL, Pant HC, Quarles RH (2002) Myelin-associated glycoprotein modulates expression and phosphorylation of neuronal cytoskeletal elements and their associated kinases. J Neurochem 81:1263–1272
12. de Waegh SM, Lee VM, Brady ST (1992) Local modulation of neurofilament phosphorylation, axonal caliber, and slow axonal transport by myelinating Schwann cells. Cell 68:451–463
13. Kirkpatrick LL, Brady ST (1994) Modulation of the axonal microtubule cytoskeleton by myelinating Schwann cells. J Neurosci 14:7440–7450

14. Veeranna, Amin ND, Ahn NG et al. (1998) Mitogen-activated protein kinases (Erk1,2) phosphorylate Lys-Ser-Pro (KSP) repeats in neurofilament proteins NF-H and NF-M. J Neurosci 18:4008–4021

15. Li BS, Veeranna, Grant P, Pant HC (1999a) Calcium influx and membrane depolarization induce phosphorylation of neurofilament (NF-M) KSP repeats in PC12 cells. Brain Res Mol Brain Res 70:84–91

16. Li BS, Veeranna, Gu J, Grant P, Pant HC (1999) Activation of mitogen-activated protein kinases (Erk1 and Erk2) cascade results in phosphorylation of NF-M tail domains in transfected NIH 3T3 cells. Eur J Biochem 262:211–217

17. Li BS, Zhang L, Gu J, Amin ND, Pant HC (2000) Integrin alpha(1) beta(1)-mediated activation of cyclin-dependent kinase 5 activity is involved in neurite outgrowth and human neurofilament protein H Lys-Ser-Pro tail domain phosphorylation. J Neurosci 20:6055–6062

18. Paglini G, Pigino G, Kunda P et al. (1998) Evidence for the participation of the neuron-specific CDK5 activator P35 during laminin-enhanced axonal growth. J Neurosci 18:9858–9869

19. Weiwad M, Werner A, Rucknagel P, Schierhorn A, Kullertz G, Fischer G (2004) Catalysis of proline-directed protein phosphorylation by peptidyl-prolyl cis/trans Isomerases. J Mol Biol 339:635–646

20. Wulf G, Finn G, Suizu F, Lu KP (2005) Phosphorylation-specific prolyl isomerization: is there an underlying theme. Nat Cell Biol 7:435–441

21. Lu PJ, Zhou XZ, Liou YC, Noel JP, Lu KP (2002) Critical role of WW domain phosphorylation in regulating phosphoserine binding activity and Pin1 function. J Biol Chem 277:2381–2384

22. Kesavapany S, Patel V, Zheng YL et al. (2007) Inhibition of Pin1 reduces glutamate-induced perikaryal accumulation of phosphorylated neurofilament-H in neurons. Mol Biol Cell 18:3645–3655

23. Davis DR, Brion JP, Couck AM et al. (1995) The phosphorylation state of the microtubule-associated protein tau as affected by glutamate, colchicine and beta-amyloid in primary rat cortical neuronal cultures. Biochem J 309:941–949

24. Brownlees J, Yates A, Bajaj NP et al. (2000) Phosphorylation of neurofilament heavy chain side-arms by stress activated protein kinase-1b/Jun N-terminal kinase-3. J Cell Sci 113:401–407

25. Shea TB, Zheng Y.-L, Ortiz D, Pant HC (2004) Cyclin-dependent kinase 5 increases perikaryal neurofilament phosphorylation and inhibits neurofilament axonal transport in response to oxidative stress. J Neurosci Res 76:795–800

26. Alvarez G, Munoz-Montano JR, Satrustegui J, Avila J, Bogonez E, Diaz-Nido J (1999) Lithium protects cultured neurons against beta-amyloid-induced neurodegeneration, FEBS Lett 453:260–264

27. Alvarez AR, Godoy JA, Mullendorff K, Olivares GH, Bronfman M (2004) Wnt-3a overcomes beta-amyloid toxicity in rat hippocampal neurons. Exp Cell Res 297:186–196

28. Maccioni RB, Otth C, Concha II, Munoz JP (2001) The protein kinase Cdk5: structural aspects, roles in neurogenesis and involvement in Alzheimer's pathology. Eur J Biochem 268:1518–1527

29. Quintanilla RA, Orellana DI, González-Billault C, Maccioni RB (2004) Interleukin-6 induces Alzheimer-type phosphorylation of tau protein by deregulating the cdk5/p35 pathway. Exp Cell Res 295:245–257

30. Ko J, Humbert S, Bronson RT et al. (2001) p35 and p39 are essential for cyclin-dependent kinase 5 function during neurodevelopment. J Neurosci 21:6758–6771

31. Shetty KT, Kaech S, Link WT et al. (1995) Molecular characterization of a neuronal-specific protein that stimulates the activity of Cdk5. J Neurochem 64:1988–1995

32. Grant P, Sharma P, Pant HC (2001) Cyclin-dependent protein kinase 5 (Cdk5) and the regulation of neurofilament metabolism. Eur J Biochem 268:1534–1546

33. Ohshima T, Ward JM, Huh CG et al. (1996) Targeted disruption of the cyclin-dependent kinase 5 gene results in abnormal corticogenesis, neuronal pathology and perinatal death. Proc Natl Acad Sci USA 93:11173–11178

34. Chae T, Kwon YT, Bronson R, Dikkes P, Li E, Tsai LH (1997) Mice lacking p35, a neuronal specific activator of Cdk5, display cortical lamination defects, seizures, and adult lethality. Neuron 18:29–42
35. Tanaka T, Veeranna, Ohshima T et al. (2001) Neuronal cyclin-dependent kinase 5 activity is critical for survival. J Neurosci 21:550–558
36. Patrick GN, Zukerberg L, Nikolic M, de la Monte S, Dikkes P, Tsai LH (1999) Conversion of p35 to p25 deregulates Cdk5 activity and promotes neurodegeneration. Nature 402:615–622
37. Bu B, Li J, Davies P, Vincent I (2002) Deregulation of cdk5, hyperphosphorylation, and cytoskeletal pathology in the Niemann-Pick type C murine model. J Neurosci 22:6515–6525
38. Whitmarsh AJ, Davis RJ (1998) Structural organization of MAP-kinase signaling modules by scaffold proteins in yeast and mammals. Trends Biochem Sci 23:481–485
39. Hunter T (2000) Signaling – 2000 and beyond. Cell 100:113–127
40. Greengard P (2001) The neurobiology of slow synaptic transmission. Science 294:1024–1030
41. Greengard P (2001) The neurobiology of dopamine signaling. Biosci Rep 21:247–269
42. Grant SG, O'Dell TJ (2001) Multiprotein complex signaling and the plasticity problem. Curr Opin Neurobiol 11:363–368
43. Wills Z, Bateman J, Korey CA, Comer A, Van Vactor D (1999) The tyrosine kinase Abl and its substrate enabled collaborate with the receptor phosphatase Dlar to control motor axon guidance. Neuron 22:301–312
44. Barford D, Das AK, Egloff MP (1998) The structure and mechanism of protein phosphatases: insights into catalysis and regulation. Annu Rev Biophys Biomol Struct 27:133–164
45. Saito T, Shima H, Osawa Y et al. (1995) Neurofilament-associated protein phosphatase 2A: its possible role in preserving neurofilaments in filamentous states. Biochemistry 34:7376–7384
46. Merrick SE, Trojanowski JQ, Lee VM (1997) Selective destruction of stable microtubules and axons by inhibitors of protein serine/threonine phosphatases in cultured human neurons. J Neurosci 17:5726–5737
47. Strack S, Westphal RS, Colbran RJ, Ebner FF, Wadzinski BE (1997) Protein serine/threonine phosphatase 1 and 2A associate with and dephosphorylate neurofilaments. Brain Res Mol Brain Res 49:15–28

Stages of Pathological Tau-Protein Processing in Alzheimer's Disease: From Soluble Aggregations to Polymerization into Insoluble Tau-PHFs

Raúl Mena and José Luna-Muñoz

Abstract Hyperphosphorylation and truncation have been proposed as key events in the abnormal tau-protein processing leading to the genesis of paired helical filaments. A recent hypothesis involving conformational changes has been emerging. However, the majority of studies have been based on the analysis of overt tangles. All the existing antibodies have been raised against normal, pathological tau protein, or intracellular tangles. It is possible that only those events occurring massively may be detected when observations are restricted to this type of structure, therefore, missing less-evident events. In general, it has been difficult to determine the early stages of tau processing in Alzheimer's disease. By the use of selected tau markers and confocal microscopy in double and triple immunolabeling and the combination with thiazin red, we have been able to determine a morphological model and the underlying molecular mechanism involved in early stages of tau-protein abnormal processing. This molecular mechanism is characterized by a hierarchical sequence of events of phosphorylation and truncation resulting in conformational misfolding along the tau molecule. We have included some speculations regarding the possible triggers of such a cascade of pathological changes of tau based on the hypothesis of truncated tau as a highly stable tau fragment with a special high affinity to bind tau monomers. Relationships between phosphorylation and the truncated mechanism are also discussed.

1 Introduction

Alzheimer's disease (AD) is a progressive degenerative disorder of insidious onset, initially characterized by memory loss and, in later stages, by severe dementia. Neurofibrillary tangles (NFTs) and dystrophic neurites, occurring in the neuropil and in neuritic plaques, are major neuropathological features of AD. Their presence in large amounts correlates with clinical dementia [1]. NFTs and dystrophic neurites are

---

R. Mena (✉) and J. Luna-Muñoz
Department of Physiology and Neurosciences,
Center of Research and Advanced Studies Mexico, DF
e-mail: rmena@fisio.cinvestav.mx

R.B. Maccioni and G. Perry (eds.) *Current Hypotheses and Research Milestones in Alzheimer's Disease*. DOI: 10.1007/978-0-387-87995-6_7,
© Springer Science + Business Media, LLC 2009

all sites of accumulation of pathological paired helical filaments (PHFs) [2]. The PHF contains the microtubule-associated protein tau as an integral structural component [3]. The assembly of the PHF appears to be central to neurofibrillary degeneration of AD neuropathology [4–6]. Tau processing leading to PHF assembly has been associated with hyperphosphorylation and truncation [6–10]. In addition to the two latter posttranslational modifications, recent reports have demonstrated that PHF-tau protein is characterized by misfoldings that are associated with early stages of tau polymerization [11–16]. NFT analysis strongly supported that these conformational changes follow a sequential pattern related to abnormally phosphorylated and truncated epitopes, which are detected by specific monoclonal antibodies (mAbs) including Alz-50, Tau-C3, Tau-66, and 423 (Table 1). In particular, one of the major molecular events involved in NFT formation appears to be associated with a truncation generated by caspase-3, an apoptosis-associated enzyme located at Asp421 of tau [17–21]. The majority of results dealing with abnormal tau processing are based on the analysis of overt tangles [21, 22]. In addition, it has not been determined whether these events are related and, if so, how they are temporally associated. Studies related to early stages of tau aggregation before its assembly into PHFs (pre-NFT stage) are not well documented [23]. Evidence shows that tau processing in pretangle cells shares similar molecular mechanisms leading to PHF formation [24]. The lack of enough studies based on the analysis of pretangle cells is mainly because of methodological

Table 1 Antibodies and recognition sites.

Antibodies	Epitope	Isotype	Reference
Alz-50	aa:5-15, 312-322. Structural conformational change	Mo IgM	[11, 12, 31]
TG-3	aa: phospho Thr231. Regional conformational change	Mo IgM	[13]
pT231	aa: phospho Thr231.	Rb IgG	
AT100	aa: phospho Ser199. Ser202, Thr205, Thr212, Ser214. Regional conformational change	Mo IgG	[36]
AT8	aa: phospho Ser199, Ser202, Thr205	Mo IgG	[7]
AD2	aa: phospho Ser396, Ser404	Mo IgG	[9]
Tau-C3	aa: Asp421 truncation	Mo IgG	[19]
423	aa: Glu391 truncation	Mo IgG	[8, 15, 34]
T-46	Aa: 404-441	Mo IgG	
M19G	Aa: firts 19	Rb IGg	

aa = aminoacids; Mo = Mouse; Rb = Rabbit; IgG = ImmunoglobulinG; IgM = Immunoglobulin M

difficulties including long-postmortem delays, stains restricted to the use of peroxidase, and the limitation of visualizing and locating these types of structures, which characteristically have a diffuse granular appearance. We have been able to settle these difficulties by the combination of double and triple immunolabeling and the use of the high-resolution confocal microscopy. In addition, we have used brain tissue from individuals with postmortem delays no longer than 6 h. We have used our methodology to investigate, in pretangle cells, whether the three molecular events showed clearly in overt tangles the hyperphosphorylation, truncation, and conformational changes that also occur [14, 25]. In addition, we addressed the issue related to a potential sequence of appearance of such molecular events in terms of a specific progression in the stages of tau processing leading to PHF assembly.

2   Stages of Abnormal Tau Processing in AD Brains

The bulk of data that have been found recently has demonstrated that the molecular mechanisms involved in tau processing in AD leading to PHF formation are driven by two physical events: aggregation and polymerization [26–29]. These two events appear to be related to time and the following of specific molecular mechanisms that include steps in which abnormal phosphorylations and truncation processes play a major role [22, 30]. These latter events would favor the generation of a series of conformational misfoldings starting from small regional changes along the tau molecule towards structural changes including the folding of the entire N-terminus along the repeated domain located along the C-terminal side of the tau protein. A clear representation of this misfolding would be the generation of the epitope detected by mAb Alz-50 [31].

3   Morphological Model of Tau-Protein Aggregation
and Polymerization in AD Brains

The state known as the "pretangle stage" presumably represents cells in which no evidence of fibrillar accumulations are observed. Morphologically, this cell population is characterized by diffuse and granular deposits of tau aggregates in a nonfibrillar state (Fig. 1a). The immunoreactive pattern displayed in pretangles is seen in the perinuclear area (*arrowhead*) and proximal processes (*arrows*). The absence of fibrillar tau clusters is confirmed because thiazin red (TR) is unable to detect pretangle cells. TR is a fluorescent dye with the property to distinguish fibrillar from nonfibrillar stages of tau aggregation [29, 32]. The next step of tau aggregation also implies sites of polymerization. This step of tangle formation appears to be represented by bead-like structures (TR positive, Fig. 1b, *arrows*) widespread throughout the perinuclear area. Characteristically, these structures are detected by TR confirming their fibrillar nature. We interpret these structures

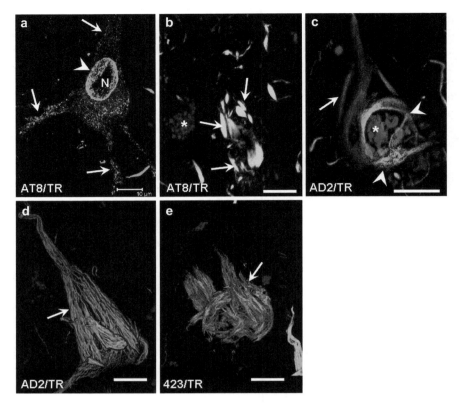

Fig. 1 Morphological model of neuronal degeneration from the pretangle state to the formation of intracellular and extracellular tangles. **a** A pretangle stage neuronal cell. Stage 1. is characterized by diffuse granular deposits throughout the perinuclear area (*arrowhead*) and proximal processes, which are undetected by thiazin red (*TR*). **b** Stage 2. corresponds to the presence of bead-like structures (paired helical filaments, PHF, nucleation sites, *arrows*), some of which are detected by TR. **c** Stage 3. is characterized by a cell having long bundles of PHFs (TR positive) covering up the nuclear and lipofucsine areas (*arrow*), the networking processing of PHF-assembled tau (*arrow*), and the putative fusion of bundles (*arrowheads*). **d** Stage 4. Intracellular neurofibrillary tangles (*NFT*) (*arrow*) having the typical flame-like appearance. **e** Stage 5. Extracellular tangles (*arrow*) whose appearance is modified by the extracellular proteolytic process. These ghost tangles only contain the PHF core, which is identified by mAb 423 and remains detected by TR [29]. Double labeling with antibodies AT8 (**a, b**), AD2 (**c, d**), and 423 (**e**) and TR. Bar = 10 μm (*See Color Plates*)

as isolated, well-formed small tangles (Fig. 1b). A following step of tau polymerization may imply the confluence of the elongating, isolated small tangles (Fig. 1c) leading to the formation of the typical flame-like tangle or intracellular tangle (I-NFT) overt NFT (Fig. 1d). Under this modeling, the extracellular tangle (E-NFT, Fig. 1e) may represent the latest step in which PHF bundles are exposed in the extracellular space. These structures are mainly constituted by a fragment of tau protein corresponding to the so-called PHF core [8, 10]. In general, the model of step-tangle formation agrees with previous studies that show evidence that E-NFTs result from I-NFTs after neuronal death [33]. We were able to define our model of tangle formation using a battery of antibodies that were able to detect I-NFTs (Table 1). E-NFTs were identified by mAb 423 [6, 10, 29, 34]. Because the patterns of immunoreactivity of all antibodies tested are shared by both pretangles and intracellular tangles, we assume that they represent, at a molecular level, different stages of the same tau-protein processing to give different morphological structures.

4   Cascade of Molecular Events that Characterize Early Stages of Tau Processing in AD

We studied the possible temporal relationships among the presence of phosphorylation, truncation, and misfolding events by the use of specific antibodies (Table 1) and confocal microscopy analysis by doing double and triple immunolabeling on pretangle neuronal cells in AD brain tissue. The sum of the results we obtained allowed us to make a model of the putative earliest stages of tau processing before their polymerization into PHF to eventually form the morphological structures, the so-called neurofibrillary tangle [32, 35]. To our knowledge, this is the first set of studies based upon the detailed analysis of neuronal cells in pretangle-stage cells in AD.

5   Hyperphosphorylation in Early Stages of Tau Processing

In this proposed model, the earliest molecular event occurring in neuronal cells vulnerable to degeneration in AD is related to a specific phosphorylation occurring at Thr^{231} and detected by the antibody pT231 (Fig. 2a). Characteristically, the distribution of this marker is located around the nucleus (*arrowhead*) and surrounding cytoplasm (*arrow*) (Fig. 2a). However, the immunoreactivity of the antibody pT231 is much stronger in the nuclear area (Fig. 2a, b). In double immunolabeling pT231, TG-3, a population of pretangle cells, is only detected by pT231 (Fig. 2a), whereas some others are double labeled with the markers (Fig. 2b). Because the pT231 marker is not related to any conformational change, our observations suggest that phosphorylation at Thr^{231} precedes the formation of the mAb TG-3 epitope and is related to a regional conformational change (Fig. 4, scheme). A later stage of tau processing would imply the formation of the mAb AT8 apitope to follow the formation of mAb AT100. We defined this sequence based on the observations of double immunuolabeling with TG3-AT8 and TG3-AT100 (Fig. 2c, d). Similar to that seen with the combination TG-3-pT231, we also distinguished two subpopulations of pretangle cells, one characterized by the presence of TG-3 in the *green channel* but with the absence of AT8 or AT100

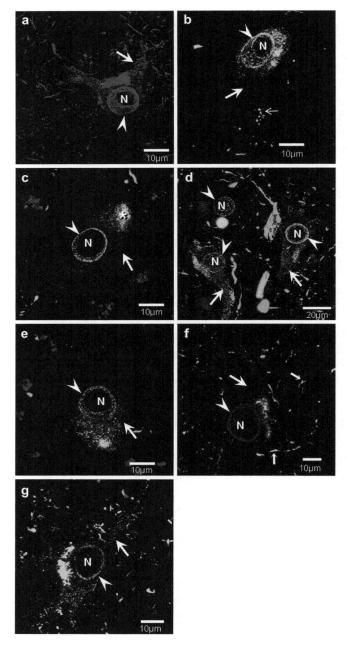

Fig. 2 a, **b** Double immunolabeling with mAb TG-3 (*green channel*) and pT231 polyclonal antibody (*red channel*). **a** mAb TG-3 was unable to detect the pretangle clearly identified by p231 in the *red channel*. In this structure pT231 immunoreactive deposits are present in the cytoplasm (*arrow*) and perinuclear area (N). **b** The two markers strongly colocated in the clusters of small dot deposits located in the near area of the three proximal processes (*short arrow*),

epitopes in the *red channel* (Fig. 2c, d, *arrows*). It is well-known that the formation of the AT100 epitope requires, at first, a sequence of phosphorylations at Ser^{199}, Ser^{202}, and Thr^{205} (conforming the mAb AT8 epitope). A second step would require further phosphorylations at Thr^{212} and Ser^{214} [36]. Although, we were unable to do double labeling with mAbs AT8 and AT100 because of methodological limitations, we can indirectly assume that mAb AT8 immunoreactivity preceded that from the mAb AT100 because the appearance of mAb TG-3 immunoreactivity precedes that of both mAbs AT8 and AT100 (Fig. 2c, d). As per analysis the putative temporal relationship of appearance of mAbs TG-3 or AT100 epitopes precede that of mAb Alz-50. We already have established that the appearance of both mAb TG-3 and AT100 epitopes occurs before formation of that of mAb Alz-50 (Fig. 2e, f) [25, 37]. However, as illustrated in Fig. 2d the appearance of the mAb TG-3 epitope also precedes that of mAb AT100 (Fig. 2d, *arrows*). When the relationship with conformational stages is considered, we also found that the regional changes along the tau molecule precede the appearance of the structural misfolding, as detected by mAb Alz-50 [25].

6 Truncation at the Early Stages of Tau Processing

To try to better understand the possible role of truncation on the early stages of tau processing, we studied the presence of truncation at Asp-421 in pretangle cells. We found that the appearance of this site of truncation seems to occur sometime between the appearance of mAb pT231 and TG-3, therefore very early during tau processing. We were able to prove this by doing triple immunolabeling with mAbs pT231, TG-3, and Tau-C3 (Fig. 3a, b), whose epitopes are the earliest formed during tau processing in pretangles. As for truncation at Glu^{391} along the C-terminus of the tau molecule, which is specifically detected by mAb 423 [10,

---

Fig. 2 (continued) perinuclear area (*arrowhead*), and cytoplasm (*arrow*). **c** mAb TG-3 (*green channel*) immunoreactive diffuse, granular deposits located in the perinuclear (*arrowhead*) and cytoplasmic areas (*arrow*) are undetected by mAb AT8 (*red channel*). Some mAb TG-3 immunoreactive granules appear to correspond to lipofuscin (*). **d** mAb TG-3 immunoreactive diffuse granular deposits located in the perinuclear area (*arrowhead*) and the cytoplasmic areas (*arrow*) are undetected by mAb AT100 in the *red channel*. **e** Cytoplasm (*arrow*) and perinuclear areas (N). A nonbearing neurofibrillary tangles (*NFT*) cell (pretangle stage) is identified by mAb TG-3 (takes the form of diffuse granular deposits). mAb Alz-50 is practically undetected in this cell. **f, g** Double immunolabeling with mAbs Alz-50 (*green channel*) and AT100 (*red channel*). **f** mAb Alz-50 was unable to detect the pretangle clearly identified by AT100 in the *red channel*. In this structure AT100 immunoreactive deposits are present in the cytoplasm (*arrow*) and perinuclear area (N). Both antibodies colocated in some neurites located in the vicinity (*short arrow*). **g** The pretangle illustrated in this figure is identified by the two markers. This colocation is more evident in the perinuclear area (N) and the small granules that are observed close to the nucleus and the contiguous process of the cell (*arrow*). *N* nucleus (*See Color Plates*)

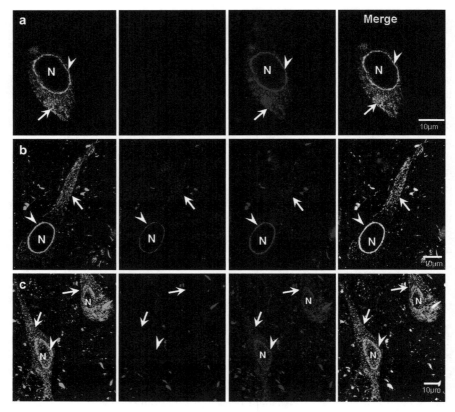

Fig. 3 a, b Triple immunolabeling with antibodies pT231 (*green channel*), Tau-C3 (*red channel*), and TG-3 (*blue channel*). Triple immunolabeling with mAbs Tau-C3 (*red channel*) and TG-3 (*blue channel*) in pretangle cells. Cells illustrated in **a, b** are detected by both pT231 and TG-3; however, Tau-C3 immunoreactivity is present in **b** but absent in **a**. Colocation between pT231 and TG-3 is more evident in the perinuclear area (*arrowhead*). **c** Triple labeling with Alz-50 (*green channel*) and the C- and N-ends of tau protein (T46, *red channel* and M19G, *blue channel*) Alz-50 immunoreactivity is present in pretangle cells (*arrows*) and perinuclear mAb Alz-50 found a double labeling between Alz-50 and N-terminus epitopes (*blue channel*) but not between Alz-50 and the C-terminus epitope (*red channel*). *N* nucleus (*See Color Plates*)

34, 38], we were unable to get a signal using immunofluorescence in pretangle cells. However, in previous studies using the peroxidase technique, immunohistochemistry, and immunoelectronmicroscopy, some evidence of mAb 423 immunolabeling was found in granular, diffuse cytoplasmic structures in AD [29]. Because of this methodological limitation, we could not analyze the presence of the mAb 423 epitope in our model of the cascade of molecular events characterizing the early stages of tau processing.

7 Modeling Early Stages of Tau Processing

7.1 Early Truncation at the C-Terminus of the Tau Molecule

From the study based on the analysis of several phosphodependent tau markers combined with those detecting conformational changes along the molecule and a truncation at Asp-421 in pretangle neuronal cells, we were able to determine a sequence of events that are temporally related and possibly represent the molecular processing that tau protein suffers, because of its soluble nature, in the formation of soluble aggregates that eventually become assembled into insoluble filaments in AD. This model is illustrated in Scheme (Fig. 4). In this model, a hypothetical tau molecule starts to undergo a cascade of molecular events in which phosphorylation at Thr^{231} may play an important role because it may favor the appearance of the first misfolding of a regional nature (TG-3, Thr^{231}). This is then followed by the progressive appearance of the mAbs AT8 (Ser^{199}, Ser^{202}, Thr^{205}), AT100 (Ser^{199}, Ser^{202}, Thr^{205}, Thr^{212}, Ser^{214}), and Alz-50 (5–15, 312–322). This cascade of events would eventually lead to the formation of tangles.

We want to emphasize three major events, which for us are of potential importance to the full interpretation of our morphological and molecular models. First, although our original model starts with a hypothetical full-length tau molecule [25], recent information has shown that a tau molecule lacking the C-terminus may be the first to be the trigger of the cascading process. For this, when we analyzed the mAb TG-3 and Alz-50 immunoreactivity related to that of the C- and N-terminal of tau molecule, we found a double labeling among TG-3, Alz-50, and N-terminus epitopes but not among Alz-50–TG-3 and the C-terminus epitope (Fig. 3c) [14]. That abnormal aggregation appears to start in a tau molecule already carrying an early C-terminus truncation. This is supported by studies showing that the cleavage of the C-terminus increases the speed of tau binding [18, 26, 27]. In addition, our results strongly support that the truncated molecule is the beginning of the proposed cascade. In addition, this truncation may be occurring after position Ser-422 (unpublished data). Second, a hypothesis proposed by Wischik et al. in 1995 [28] introduced the challenging concept that the tau assembly is a consequence of the presence of a "primer" or a highly stable tau fragment (a tau species truncated at Glu^{391} with a characteristic high affinity to intact tau molecules), which binds tau molecules to become a stable complex. By means of a subsequent truncation in an autocatalytic process involving binding–truncation–binding this is triggered exponentially. This hypothesis is supported by recent studies in vitro, which have demonstrated that the truncated tau protein has exponentially higher kinetics of binding to tau than the intact tau molecule [18, 26, 27, 39]. In addition, confocal and immunoelectronmicroscopy studies have provided strong evidence that mAb 423, which specifically identifies the truncated tau at the Glu^{391} and in the so-called PHF "core," is present in a diffuse, granular, and amorphous material in the cytoplasm of neuronal cells in AD brains. This bulk of evidence allowed us to suggest that the dynamic tau processing, which we have described and involving a specific cascade

Fig. 4 Schematic drawing illustrating the cascade of events characterizing the early stages of tau processing in pretangle neuronal cells. The scheme starts from the hypothetical appearance of a stable tau fragment corresponding to the minimal core, which is characteristically truncated at the position Glu[391] of the C-terminus. We add a question mark because this step is not yet clearly defined. The actual cascade of molecular events starts from either an intact or a C-terminally truncated tau molecule (**a**). From this, a sequence of phosphorylations starting with the addition of a phosphate group at Thr[231] (**b**) will determine the appearance of first two regional conformational changes found in the molecule and detected by mAbs TG-3 (**d**) and AT100 (**f**). After this misfolding the structural change of the molecule, as detected by mAb Alz-50 (**g**), will then occur just before the appearance of the first truncation in the position Asp[421] at the C-terminus and detected by mAb Tau-C3. (**c**). This cascade of specific molecular events will lead to the eventual formation of tangles and the ultimate exposure of the Pronase-resistant paired helical filaments (*PHF*) or PHF core as demonstrated elsewhere [6, 33]

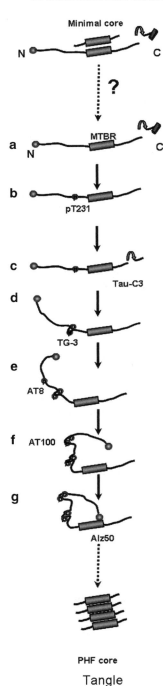

of phosphorylation, may result from the presence of the minimal core (truncated tau at Glu^{391}). This fragment which would initially bind either intact or C-terminus-truncated tau monomers, thus developing the sequence of events already described in our scheme (Fig. 4). This processing would lead to the eventual formation of the pronase-resistant PHF fractions [8, 10] and their ultimate exposure in the ghost tangle after neuronal death [6, 16, 22, 29]. Third, as for states of solubility of tau filaments, we want to speculate about the differences between aggregation and polymerization. Our model is based on the analysis of the immunoreactivity patterns shown by several antibodies (Table 1), all of which characteristically take the form of diffuse and granular deposits in pretangle neuronal cells consistent with a nonfibrillar state of tau-molecule aggregates. These deposits are also characteristically undetected by TR, supporting the absence of β-pleated sheet structures (assembled PHFs). As for polymerization as a fibrillar state of tau aggregates, the small tangles that are described in our morphological model would represent the first bundles of PHF, which are progressively accumulated in the neuronal cytoplasm. Characteristically, these bundles are strongly detected by TR; therefore, they contain the β-pleated sheet structure.

8 Conclusion

Whatever are the underlying molecular mechanisms directly involved in the abnormal tau processing of tau proteins in AD, it still remains to determine the place where this process may occur. Some evidence suggests that some membranous organelles, including mitochondria, lysosomes, endoplasmic reticulum, and the nucleus [29, 40], may be associated with this cascade of events affecting charges and conformations of tau via phosphorylation and truncations. It is relevant to try to find the precise location within the cytoplasmic milieu and to determine whether the same type of cytoplasmic processing is also occurring in the neuronal processes where these pathological mechanisms have not yet been studied.

Acknowledgments Authors express their gratitude to Dr. P. Davies (Albert Einstein College of Medicine, Bronx, NY, USA) for the generous gift of mAbs TG-3, Alz-50, and MC1, Mr. José L. Fernández for handling of the brain tissue, and Ms. Maricarmen De Lorenz for her secretarial assistance. This work was financially supported by CONACyT grants, No. 47630 (to R.M.). Thanks to Dr. Ellis Glazier for editing the English-language text.

References

1. Arriagada PV, Growdon JH, Hedley-Whyte ET, Hyman BT (1992) Neurofibrillary tangles but not senile plaques parallel duration and severity of Alzheimer's disease. Neurology 42: 631–639

2. Kidd M (1963) Paired helical filaments in electron microscopy of Alzheimer's disease. Nature 197: 192–193

3. Kondo J, Honda T, Mori H, et al. (1988) The carboxyl third of tau is tightly bound to paired helical filaments. Neuron 1: 827–834

4. Braak E, Braak H, Mandelkow EM (1994) A sequence of cytoskeleton changes related to the formation of neurofibrillary tangles and neuropil threads. Acta Neuropathol 87: 554–567

5. Braak H, Braak E (1991) Neuropathological stageing of Alzheimer-related changes. Acta Neuropathol 82: 239–259

6. Mena R, Edwards PC, Harrington CR, et al. (1996) Staging the pathological assembly of truncated tau protein into paired helical filaments in Alzheimer's disease. Acta Neuropathol 91: 633–641

7. Goedert M, Jakes R, Crowther RA, et al. (1993) The abnormal phosphorylation of tau protein at Ser-202 in Alzheimer disease recapitulates phosphorylation during development. Proc Natl Acad Sci USA 90: 5066–5070

8. Novak M, Kabat J, Wischik CM (1993) Molecular characterization of the minimal protease resistant tau unit of the Alzheimer's disease paired helical filament. EMBO J 12: 365–370

9. Trojanowski JQ, Schmidt ML, Shin RW, et al. (1993) Altered tau and neurofilament proteins in neuro-degenerative diseases: diagnostic implications for Alzheimer's disease and Lewy body dementias. Brain Pathol 3: 45–54

10. Wischik CM, Novak M, Edwards PC, et al. (1988) Structural characterization of the core of the paired helical filament of Alzheimer disease. Proc Natl Acad Sci USA 85: 4884–4888

11. Jicha GA, Berenfeld B, Davies P (1999) Sequence requirements for formation of conformational variants of tau similar to those found in Alzheimer's disease. J Neurosci Res 55: 713–723

12. Jicha GA, Bowser R, Kazam IG, et al. (1997) Alz-50 and MC-1, a new monoclonal antibody raised to paired helical filaments, recognize conformational epitopes on recombinant tau. J Neurosci Res 48: 128–132

13. Jicha GA, Lane E, Vincent I, et al. (1997) A conformation- and phosphorylation-dependent antibody recognizing the paired helical filaments of Alzheimer's disease. J Neurochem 69: 2087–2095

14. Luna-Muñoz J, García-Sierra F, Falcón V, et al. (2005) Regional conformational change involving phosphorylation of tau protein at the Thr231, precedes the structural change detected by Alz-50 antibody in Alzheimer's disease. J Alzheimers Dis 8: 29–41

15. Skrabana R, Kontsek P, Mederlyova A, et al. (2004) Folding of Alzheimer's core PHF subunit revealed by monoclonal antibody 423. FEBS Lett 568: 178–182

16. Mena R, Wischik CM, Novak M, et al. (1991) A progressive deposition of paired helical filaments (PHF) in the brain characterizes the evolution of dementia in Alzheimer's disease. An immunocytochemical study with a monoclonal antibody against the PHF core. J Neuropathol Exp Neurol 50: 474–490

17. Fasulo L, Ugolini G, Visintin M, et al. (2000) The neuronal microtubule-associated protein tau is a substrate for caspase-3 and an effector of apoptosis. J Neurochem 75: 624–633

18. Gamblin TC, Berry RW, Binder LI (2003) Modeling tau polymerization in vitro: a review and synthesis. Biochemistry 42: 15009–15017

19. Gamblin TC, Chen F, Zambrano A, et al. (2003) Caspase cleavage of tau: linking amyloid and neurofibrillary tangles in Alzheimer's disease. Proc Natl Acad Sci USA 100: 10032–10037

20. Guillozet-Bongaarts AL, Cahill ME, Cryns VL, et al. (2006) Pseudophosphorylation of tau at serine 422 inhibits caspase cleavage: in vitro evidence and implications for tangle formation in vivo. J Neurochem 97: 1005–1014

21. Guillozet-Bongaarts AL, Garcia-Sierra F, Reynolds MR, et al. (2005) Tau truncation during neurofibrillary tangle evolution in Alzheimer's disease. Neurobiol Aging 26: 1015–1022

22. García-Sierra F, Ghoshal N, Quinn B, et al. (2003) Conformational changes and truncation of tau protein during tangle evolution in Alzheimer's disease. J Alzheimers Dis 5: 65–77

23. Bancher C, Brunner C, Lassmann H, et al. (1989) Accumulation of abnormally phosphorylated tau precedes the formation of neurofibrillary tangles in Alzheimer's disease. Brain Res 477: 90–99

24. Augustinack JC, Schneider A, Mandelkow EM, et al. (2002) Specific tau phosphorylation sites correlate with severity of neuronal cytopathology in Alzheimer's disease. Acta Neuropathol 103: 26–35

25. Luna-Muñoz J, Chávez-Macías L, García-Sierra F, et al. (2007) Earliest stages of tau conformational changes are related to the appearance of a sequence of specific phospho-dependent tau epitopes in Alzheimer's disease. J Alzheimers Dis 12: 365–375

26. Berry RW, Abraha A, Lagalwar S, et al. (2003) Inhibition of tau polymerization by its carboxy-terminal caspase cleavage fragment. Biochemistry 42: 8325–8331

27. Gamblin TC, Berry RW, Binder LI (2003) Tau polymerization: role of the amino terminus. Biochemistry 42: 2252–2257

28. Wischik CM, Edwards PC, Lai RY, et al. (1995) Quantitative analysis of tau protein in paired helical filament preparations: implications for the role of tau protein phosphorylation in PHF assembly in Alzheimer's disease. Neurobiol Aging 16: 409–417

29. Mena R, Edwards P, Pérez-Olvera O, et al. (1995) Monitoring pathological assembly of tau and beta-amyloid proteins in Alzheimer's disease. Acta Neuropathol 89: 50–56

30. Mondragón-Rodríguez S, Basurto-Islas G, Santa-Maria I, et al. (2008) Cleavage and conformational changes of tau protein follow phosphorylation during Alzheimer's disease. Int J Exp Pathol 89: 81–90

31. Carmel G, Mager EM, Binder LI, et al. (1996) The structural basis of monoclonal antibody Alz50s selectivity for Alzheimer's disease pathology. J Biol Chem 271: 32789–32795

32. Uchihara T, Nakamura A, Yamazaki M, et al. (2001) Evolution from pretangle neurons to neurofibrillary tangles monitored by thiazin red combined with Gallyas method and double immunofluorescence. Acta Neuropathol 101: 535–539

33. García-Sierra F, Wischik CM, Harrington CR, et al. (2001) Accumulation of C-terminally truncated tau protein associated with vulnerability of the perforant pathway in early stages of neurofibrillary pathology in Alzheimer's disease. J Chem Neuroanat 22: 65–77

34. Novak M (1994) Truncated tau protein as a new marker for Alzheimer's disease. Acta Virol 38: 173–189

35. Vincent I, Zheng JH, Dickson DW, et al. (1998) Mitotic phosphoepitopes precede paired helical filaments in Alzheimer's disease. Neurobiol Aging 19: 287–296

36. Zheng-Fischhöfer Q, Biernat J, Mandelkow EM, et al. (1998) Sequential phosphorylation of Tau by glycogen synthase kinase-3beta and protein kinase A at Thr212 and Ser214 generates the Alzheimer-specific epitope of antibody AT100 and requires a paired-helical-filament-like conformation. Eur J Biochem 252: 542–552

37. Weaver CL, Espinoza M, Kress Y, et al. (2000) Conformational change as one of the earliest alterations of tau in Alzheimer's disease. Neurobiol Aging 21: 719–727

38. Novak M, Wischik CM, Edwards P, et al. (1989) Characterisation of the first monoclonal antibody against the pronase resistant core of the Alzheimer PHF. Prog Clin Biol Res 317: 755–761

39. Abraha A, Ghoshal N, Gamblin TC, et al. (2000) C-terminal inhibition of tau assembly in vitro and in Alzheimer's disease. J Cell Sci 113 Pt. 21: 3737–3745

40. Galván M, David JP, Delacourte A, et al. (2001) Sequence of neurofibrillary changes in aging and Alzheimer's disease: a confocal study with phospho-tau antibody, AD2. J Alzheimers Dis 3: 417–425

Plasma Membrane-Associated PHF-Core Could be the Trigger for Tau Aggregation in Alzheimer's Disease

Karla I. Lira-De León, Martha A. De Anda-Hernández, Victoria Campos-Peña, and Marco A. Meraz-Ríos

Abstract In the present analysis, we discuss the possible role of tau in neurodegeneration, when its intracellular normal location is altered. In order to validate our hypothesis, we used two chimeric constructs: the first one is the fusion of the membrane anchorage signal of interferon-γ receptor α-chain to the paired helical filament (PHF)-core fragment of 94 amino acids (aa) ending in glutamic 391 (IFNγR-NMF) and the second one is the PHF-core linked to the 100aa of the C-terminus of the amyloid-β protein precursor (AβPP-C100), which contains the transmembranal domain and was named Aβ-TMD-NMF. Both constructs showed thiazine red positive signal when they were cotransfected with tau441, indicating presence of β-sheet structures in cos7 cultures. In addition, when IFNγR-NMF construction was introduced in neural precursor cells primary cultures, the appearances of β-sheet structures by thiazine red signal were observed. These observations indicate that when the chimeric construct interacts either with endogenous or exogenous tau, it is capable of inducing the formation of abnormal polymerized tau. These data support the phenomena observed by Wischick et al. in vitro, and corroborate the idea of the importance of tau interacting with the membrane and the possible role as a nucleation center for PHF formation, suggesting that the abnormal localization of tau in the plasma membrane could be an early step in Alzheimer's disease.

Keywords Alzheimer's disease, Tau, β-Sheet structures, Mislocalization, Truncation, PHFs

Abbreviations AD: Alzheimer's disease; AβPP: amyloid-β protein precursor; NPC: neural precursor cells; PHFs: paired helical filaments

M.A. Meraz-Rios ($\boxtimes$)
mmeraz@cinvestav.mx

R.B. Maccioni and G. Perry (eds.) *Current Hypotheses and Research Milestones in Alzheimer's Disease*. DOI: 10.1007/978-0-387-87995-6_8,
© Springer Science + Business Media, LLC 2009

1 Introduction

Alzheimer's disease (AD), the most common dementia, is characterized by two patho-logical protein deposits in the brain; the amyloid plaques composed by fibrous assem-blies of the amyloid-β (Aβ) peptide derivative of the membrane amyloid-β protein precursor (AβPP) and the neurofibrillary tangles, which are bundles of paired helical filaments (PHFs) whose main constituent is the microtubule-associated protein, tau. These aggregates are toxic to neurons, either causing some toxic signaling defect (Aβ deposits) or obstructing the cell interior (tau deposits). Both deposits have been reported as responsible for neuronal degeneration. Therefore, it is very important to understand the factors that promote the abnormal aggregation of Aβ and tau protein [1, 2].

The Aβ-peptide is partly hydrophobic which interacts across and along peptide strands favoring the formation of stable β-sheets. Otherwise, cytosolic tau is a very hydrophilic and highly soluble protein; thus, it shows a poor tendency to aggregate in physiological buffer conditions, and the formation of aggregates is very slow. This explains the unfolded nature of tau; however, it does not explain the abnormal aggregation in AD [3].

2 Plasmatic Membrane Elements and PHF Formation

As a result of many experiments performed in vitro, the following events for self-aggrega-tion of tau have been defined; (a) the interaction among molecules with negative charges like fatty acids, polyanionic molecules, and RNA may increase tau aggregation, (b) the microtubule-binding domain of tau is the core of this assembly [4], and (c) oxidation, hyperphosphorylation, or truncation of tau is required to accelerate its addition [5, 6].

Although these substances in vitro promote tau polymerization, in vivo, only anionic lipids or lipid-derived substances (such as fatty acids) have been implicated in AD. Gray et al. [7] observed in biopsies from AD brains that the PHFs appear to originate from the surface of a cytomembrane. This observation correlates with the normal interaction of tau with the neural cytoplasmic membrane through its N-terminal domain, and this association is regulated by phosphorylation of tau [8, 9]. Tau in situ may be associated directly with the plasma membrane phospholipids or with additional membrane-associated protein(s), and participate in signal trans-duction mechanisms [10]. At the membrane, tau interacts with the Src-family nonreceptor tyrosine kinase, mediated by the proline-rich region of tau and the SH3 domain of Fyn or Scr [11]. As a result of these findings, it has been suggested that a component of the membrane can nucleate tau assembly.

3 Catalytic Aggregation Model of tau

Wischick et al. (1988) observed by electron microscopy of negatively stained speci-mens that PHFs had a different morphology when they were treated with pronase compared with untreated filaments. These data reveal that untreated filaments have a fuzzy outer covering, which is removed by protease [12]. Pronase treatment removes approximately 17% of the material that corresponds to the larger N- and smaller C-terminal region of the tau polypeptide. This pronase-resistant portion was named the PHF-core [12].

The minimal protease-resistant tau unit is 93–95 residues long, the equivalent of three repeats, but is 14–16 residues out of phase with respect to the maximum homology organization of the repeat region. This PHF-core has truncated tau at Glu391 which is produced by endogenous proteases as one of the events leading to PHF assembly, or alternatively that it reflects the action of proteases on a partially assembled precursor [13]. Self-aggregation confers proteolytic stability on a short segment that excludes the N-terminal half of the tau molecule and reproduces the characteristic Glu391 truncation at the C-terminus [14].

Once this was noticed, the pronase-stripped filaments were used as immunogen to raise a monoclonal antibody (mAb) against the PHF-core which reacts with pronase-stripped filaments much more strongly than it does with unstripped ones [12]. The mAb423 reacts with epitopes present in the 9.5 and 12 kDa fragments and recognizes a specific C-terminal cleavage site at Glu391. The selectivity of mAb423 has been useful in several histological, ultrastructural, and biochemical studies of neurofibrillary pathology in AD [13].

It has been proposed that aggregation involves antiparallel dimerization [14]. According to this hypothesis, the folding of the repeat region within the PHF-core does not appear to reflect the same organization as the full-length tau molecule [13]. A possible mechanism of tau sequestration and PHF formation has been suggested: the core PHF-tau fragment binds full-length tau with an affinity comparable to a strong antibody–antigen interaction, and those regions of the molecule outside the core fragment do not enhance the binding coefficient. Self-aggregation confers proteolytic stability to a short segment that excludes the N-terminal half of the molecule and reproduces the characteristic Glu391 truncation at the C-terminus. Truncated aggregates generated in vitro retain the capacity to propagate tau capture and to seed the further accumulation of truncated tau in the presence of proteases. Once the process has started, continued generation of the truncated tau fragment in the solid phase has the intrinsic capacity for propagation of tau capture in the presence of proteases. Then, it can be concluded that PHFs require a pathological tau–tau binding interaction through the repeat domain, and only in this region of the molecule, and that this interaction is required to maintain proteolytic stability of the PHF-core [14].

4   Cell Models and Chimeric Constructs of Tau

Observations of Wischick et al. lead to the idea of the importance of tau interacting with the membrane and the possible role of the site of interaction as a nucleation center for PHF formation [14]. Therefore, we decided to promote tau interaction with cytoplasmic membranes through a chimeric construct that could promote the aggregation of cytoplasmic tau and lead to the formation of β-sheet structures characteristics of PHF.

With this purpose, we used the sequence of full-length human tau (isoform 441aa) as well as the PHF-core (297–391aa) to transfect COS7 cells and neural precursor cells (NPCs) primary cultures. These sequences were subcloned in pcDNA3 vector for use in COS7 cells and sequenced; their expression was confirmed by Western blot assays (data not showed). With a subcloning strategy, the PHF-core was linked in the N-terminal region to the α-chain of the interferon-γ receptor (1–271aa) lacking the intracellular domain (IFNγR-NMF) and a construct where the PHF-core was linked to the 100aa of the C-terminus of AβPP (C100), which contains the transmembranal domain and was named Aβ-TMD-NMF (Fig. 1a). COS7 cells were stably transfected with (a) full-length tau, (b) IFNγR-NMF, and (c) Aβ-TMD-NMF or with one chimerical construct and full-length tau (Fig. 1b).

Confocal microscopy revealed that the full-length tau induces neither any evident morphological change nor the presence of β-sheet structures, and stable transfection of tau showed a classical pattern of cytoplasmic proteins. Also, the stable transfection with IFNγR-NMF shows a classical membrane pattern and no formation of β-sheet structures, revealed by thiazin red (TR) staining (Fig. 1b). When we cotransfected full-length tau and IFNγR-NMF, we observed the presence of β-sheet structures after TR staining (Fig. 1c). β-Sheet-pleated structure formation was present in approximately 60% of the cotransfected cells. The same result was observed with cotransfected full-length tau and Aβ-TMD-NMF (Fig. 1c). These results suggest that PHF-core anchored to plasma membrane has the capability to induce an abnormal redistribution and processing of full-length tau into TR-positive structures [15].

In order to approach the microenvironment in which the abnormal aggregation of tau, a characteristic of AD, occurs, we used NPC primary cultures (Fig. 2). IFNγR-NMF and tau441 constructs (Fig. 1a) used in COS7 cells were employed in a system of replication incompetent retroviral vectors [16]. With these constructs, we infected NPCs and evaluated the possible generation of β-structures. As expected, the expression of IFNγR-NMF was observed on the plasma membrane and stained with mAb423, which recognizes the truncation in tau at Glu391 (Fig. 2a). Mock cells transduced with empty virus shows no staining with this antibody. Also this construct showed colocalization with endogenous tau, which was stained with polyclonal antibody tau Ab3 that recognized the C-terminus of tau (Fig. 2a).

Fig. 1 (continued) (b) PHF-core as reported (12), (c) PHF-core anchored to the α-chain of interferon-γ receptor, and (d) PHF-core anchored to APPC100. **b** Effect of stable transfection of full-length tau in COS7 cells (a), IFNγR-NMF (b), and Aβ-TMD-NMF (c), using FITC as secondary antibody; any of them promotes the presence of β-sheet structures. **c** Cotransfection of full-length tau with each chimerical construct (IFNγR-NMF and Aβ-TMD-NMF, respectively) generated β-sheet structures revealed by TR; tau detection made by HT7 recognized human tau; and amyloid-β was detected with antiamyloid-β antibody that recognizes Aβ 1–17 [15] (*see Color Plates*)

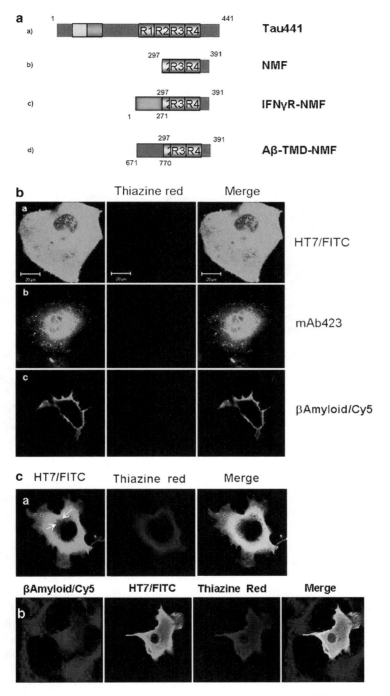

Fig. 1 PHF-core anchored to plasma membrane induce the presence of β-sheet structures in COS7 cells culture. **a** Schematic representation of constructs used to transfect COS7 cell culture or transduce primary cultures of neural precursor cells: (a) human full-length (isoform 441),

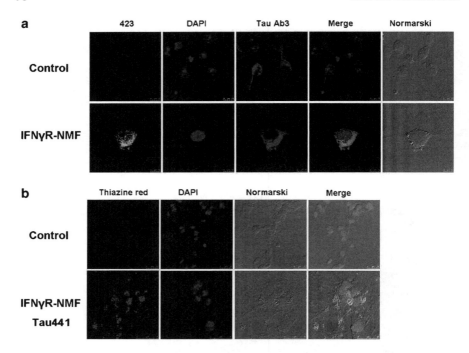

Fig. 2 Generation of β-sheet structures in neural precursor cell (*NPC*) induced by IFNγR-NMF. **a** Immunofluorescence staining of endogeneous tau protein (antibody tau Ab3) showed a partial colocalization with IFNγR-NMF expression (mAb423). **b** Cells that express IFNγR-NMF and tau441 showed positive staining of β-sheet structures by TR (*See Color Plates*)

Interestingly when we coexpressed IFNγR-NMF and full-length tau (tau441), we observed positive staining with TR (Fig. 2b), suggesting the presence of β-sheet structures, which is the characteristic of PHFs.

These data suggest that the interaction of tau with the plasma membrane may participate in the pathological pathway of tau in AD. On normal conditions, tau is able to stabilize microtubules in a phosphorylation-dependent manner. This phosphorylation also regulates tau interaction with plasma membrane through membrane proteins (such as Fyn, [11]) or directly with phospholipids [10] (Fig. 3a). In AD, hyperphosphorylated tau could be present and unable to bind to microtubules, and as a consequence, the accumulated tau would show a conformational change promoting truncation in Asp421 and then in Glu391, as well as its truncation in the N-terminus [17]. Truncated tau is capable of promoting self-aggregation through the microtubule-repeat region in an antiparallel way. This aggregation leads to oligomer formation [18] that could lead to interaction with tau membrane and become a nucleation center, resulting in possible PHF formation (Fig. 3b).

In conclusion, a chimeric tau protein anchorage to plasma membrane could capture cytoplasmic full-length tau and promote the generation of β-sheet structures.

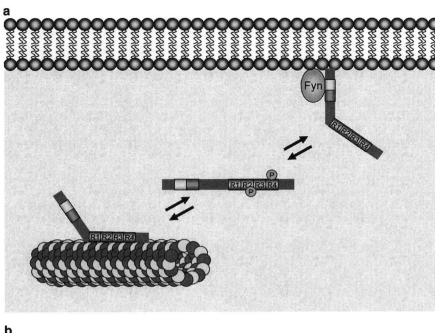

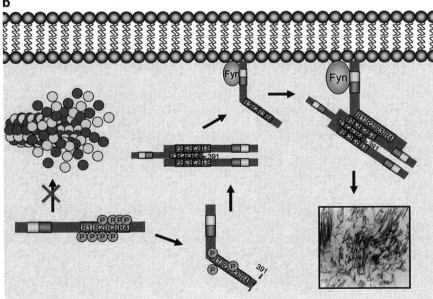

Fig. 3 Hypothetical model of PHF assembly. See text for details (*See Color Plates*)

References

1. Friedhoff P, von Bergen M, Mandelkow EM et al. (1998) A nucleated assembly mechanism of Alzheimer paired helical filaments. *Proc Natl Acad Sci USA* 95:15712–15717
2. von Bergen M, Friedhoff P, Biernat J et al. (2000) Assembly of τ protein into Alzheimer paired helical filaments depends on a local sequence motif (^{306}VQIVYK^{311}) forming β structure. *Proc Natl Acad Sci USA* 97:5129–5134
3. Mandelkow E, von Bergen M, Biernat J and Mandelkow EM (2007) Structural principles of tau and the paired helical filaments of Alzheimer's disease. *Brain Pathol* 17:83–90
4. Skrabana R, Sevcik J, Novak M (2006) Intrinsically disordered proteins in the neurodegenerative processes: formation of tau protein paired helical filaments and their analysis. *Cell Mol Neurobiol* 26:1085–1097
5. King ME, Ahuja V, Binder LI, Kuret J (1999) Ligand-dependent tau filament formation: implications for Alzheimer's disease progression. *Biochemistry* 38:14851–14859
6. Alonso A, Zaidi T, Novak M et al. (2001) Hyperphosphorylation induces self-assembly of tau into tangles of paired helical filaments/straight filaments. *Proc Natl Acad Sci USA* 98:6923–6928
7. Gray EG, Paula-Barbosa M, Roher A (1987) Alzheimer's disease: paired helical filaments and cytomembranes. *Neuropathol Appl Neurobiol* 13:91–110
8. Brandt R, Leger J, Lee G (1995) Interaction of tau with the neural plasma membrane mediated by tau's amino-terminal projection domain. *J Cell Biol* 131:1327–1340
9. Ekinci FJ, Shea TB (2000) Phosphorylation of tau alters its association with the plasma membrane. *Cell Mol Neurobiol* 20:497–508
10. Shea TB (1997) Phospholipids alter tau conformation, phosphorylation, proteolysis, and association with microtubules: implication for tau function under normal and degenerative conditions. *J Neurosci Res* 50:114–122
11. Lee G (2005) Tau and src family tyrosine kinases. *Biochim Biophys Acta* 1739:323–330
12. Wischik CM, Novak M, Edwards PC et al. (1988) Structural characterization of the core of the paired helical filament of Alzheimer disease. *Proc Natl Acad Sci USA* 85:4884–4888
13. Novak M, Kabat J, Wischik CM (1993) Molecular characterization of the minimal protease resistant tau unit of the Alzheimer's disease paired helical filament. *EMBO J* 12:365–370
14. Wischik CM, Edwards PC, Lai RY et al. (1996) Selective inhibition of Alzheimer disease-like tau aggregation by phenothiazines. *Proc Natl Acad Sci USA* 93:11213–11218
15. Campos-Peña V, Mena R, Franco Lira M, Meraz Rios MA (2006) The Plasmatic Expression of PHF-Core Induce the Presence of Structure Recognized by TR. 10th International Conference on Alzheimer's Disease and Related Disorders. MEDIMOND International Proceedings 467–474
16. Cárdenas-Aguayo MC, Santa-Olalla J, Baizabal JM et al. (2003) Growth factor deprivation induces an alternative non-apoptotic death mechanism that is inhibited by Bcl-2 in cells derived from neural precursor cells. *J Hematother Stem Cells Res* 12:735–748
17. Binder LI, Guillozet-Bongaarts AL, Garcia-Sierra F, Berry RW (2005) Tau, tangles, and Alzheimer's disease. *Biochim Biophys Acta* 1739:216–223
18. Berger Z, Roder H, Hanna A et al. (2007) Accumulation of pathological tau species and memory loss in a conditional model of tauopathy. *J Neurosci* 27:3650–3662

Oxidative Stress Hypothesis

Neurofibrillary Tangle Formation as a Protective Response to Oxidative Stress in Alzheimer's Disease

Akihiko Nunomura, Atsushi Takeda, Paula I. Moreira, Rudy J. Castellani, Hyoung-gon Lee, Xiongwei Zhu, Mark A. Smith, and George Perry

Abstract Neurofibrillary tangles (NFTs) composed of hyperphosphorylated tau are major hallmarks of Alzheimer's disease (AD). Because the formation of NFTs reflects a hierarchy of neuron al vulnerability and their distribution parallels disease severity, NFTs formation has been suspected to play a major role in the disease pathogenesis. However, theoretically, either pathogenic alterations of the disease or protective responses to the disease pathogenesis can be observed according to the hierarchy of the vulnerability. Indeed, the majority of neuronal death in AD likely occurs without the process of NFT formation and neurons may live for decades with NFTs. More important, there is a growing body of evidence suggesting that tau phosphorylation and conformational changes are inducible by oxidative insults and the neuronal oxidative damage in AD is actually alleviated through the process of NFT formation. In line with recent evidence that neuronal cellular inclusions represent a protective function, rather than being initiators or accelerators of disease pathogenesis, we suspect that the NFTs function as a cytoprotective response especially a primary line of antioxidant defense. An involvement of tau phosphorylation in the insulin-like signaling pathway affecting organism longevity implicates an essential link between NFT formation and an adaptation under oxidative stress in age-associated neurodegeneration.

Keywords Aging; Alzheimer's disease; Free radical; Neurofibrillary tangles; Oxidative stress; Phosphorylation; Tau,

A. Nunomura (✉), A. Takeda, P.I. Moreira, R.J. Castellani, H.-gon Lee, X. Zhu, M.A. Smith, and G. Perry
Department of Neuropsychiatry, Interdisciplinary Graduate School of Medicine and Engineering, University of Yamanashi, 1110 Shimokato,
Chuo, Yamanashi 409-3898, Japan
e-mail: anunomura@yamanashi.ac.jp

G. Perry (✉)
College of Sciences, University of Texas at San Antonio, 6900 North Loop,
1604 West, San Antonio, TX 78249
e-mail: george.perry@utsa.edu

R.B. Maccioni and G. Perry (eds.) *Current Hypotheses and Research Milestones in Alzheimer's Disease*. DOI: 10.1007/978-0-387-87995-6_9,
© Springer Science + Business Media, LLC 2009

1   Introduction

Amyloid-β (Aβ) and a microtubule-associated protein tau that are major constituents of senile plaques and neurofibrillary tangles (NFTs) in Alzheimer's disease (AD), respectively, are among the best-studied proteins in all of neurobiology and figure centrally into much of the research dedicated to AD. While not surprising since the pathological diagnosis of AD is dependent upon the quantity of Aβ and tau deposits within cortical gray matter [1, 2], we suggest that this strict linkage of diagnostic and mechanistic views is misleading, particularly in the case of neurodegenerative diseases. Neuropathological changes in subjects with dementia are, by definition, end-stage phenomena. Although such changes allow case characterization and lend themselves to disease classification and modeling, the lesions themselves are not etiological. They are certainly pathognomonic but not necessarily pathogenic. Theoretically, either pathogenic alterations of the disease or protective responses to the disease pathogenesis can be observed according to the hierarchy of the neuronal vulnerability, of which the latter is the case in neurodegenerative diseases. This short chapter focuses on tau pathology and the process of NFT formation in AD and its involvement in a compensatory response against oxidative stress.

2   Tau Pathology in Normal Aging

In the population of normal aging, prevalence of NFTs is increased with advancing age [3]. Even in the twenties, more than 10% of the population exhibits NFTs of Braak stages I and II characterized by entorhinal NFTs. In the forties, about 40% of the population possesses the entorhinal NFTs, whereas the appearance of NFTs compatible with Braak stages III and IV characterized by limbic NFTs or Braak stages V and VI characterized by neocortical NFTs in normal population starts only after age 50 [3]. As for the elderly subjects, NFTs are present in a considerable percentage of brains of cognitively normal. A study investigating autopsied subjects aged between 69 and 100 who were cognitively normal revealed that 27% of subjects are in Braak stages III and IV (limbic NFTs) and 10% of subjects are in Braak stages V and VI (neocortical NFTs) [4], which informs us that even mature and abundant tau pathology indistinguishable from AD brain often fails to cause cognitive dysfunction in the elderly.

3   Tau Pathology in AD and Other Tauopathies

In AD, in contrast with a poor correlation between Aβ plaque density and neuronal loss or disease severity, NFT density correlates with neuronal loss and clinical severity [5–7]. However, the amount of neuronal loss largely exceeds the amount of NFTs [6, 8], which strongly suggests that most of the neurons in AD die via non-NFT formation.

Because a microtubule-associated protein tau physiologically has a role in maintaining stability of microtubules, tau alterations are believed to cause disassembly of microtubules and subsequently compromise microtubule function, resulting in a decline in axonal or dendritic transport. However, an ultrastructural analysis of AD brain sample demonstrated that a reduction in number and total length of microtubules seen in pyramidal neurons in AD was unrelated to the presence of NFTs [9]. Also, it has been shown that although unpolymerized hyperphosphorylated tau in the cytosol can sequester normal functional tau and causes microtubule disassembly [10], polymerized tau in the form of NFTs loses this ability [11]. In AD, neurons may therefore promote NFT formation to protect function of normal cytosolic tau, thereby allowing neurons to survive longer [12].

Indeed, neurons with NFTs are estimated to be able to survive for decades [13], which suggests that NFTs themselves are not obligatory for neuronal death in AD. There may be two pathways to neuronal death: one is accompanied by NFT formation in which neurons slowly degenerate, and the other is through non-NFT formation in which neurons die quickly [14]. In other words, vulnerable neurons under certain etiological insults in AD can live longer due to compensatory cellular mechanisms associated with NFT formation [15].

Tau is the major component of the intracellular filamentous deposits that define not only AD but also a number of neurodegenerative diseases known collectively as tauopathies. They include AD, progressive supranuclear palsy, corticobasal degeneration, Pick's disease, and argyrophilic grain disease, as well as the inherited frontotemporal dementia and parkinsonism linked to chromosome 17 (FTDP-17) [16]. When tau load in the frontal cortex is compared by image analysis of immunohistochemically stained sections using the phospho-dependent antibodies in patients with FTDP-17, sporadic FTLD with Pick bodies, and early-onset AD, the amount of tau in FTDP-17 and sporadic FTLD with Pick bodies is significantly less than that in early-onset AD [17]. This observation is somewhat paradoxical given the prevailing view that mutations in the tau gene are the root cause of FTDP-17, whereas tau pathology in AD is considered to be more downstream to Aβ pathology. Also, discrepancy of the amount of tau load and neuronal damage can be mentioned between FTLD and AD because there is greater tissue loss in FTLD than in AD. These findings suggest the possibility that aggregation of tau protein represents cellular adaptive response in the tauopathies.

Recently established animal models of tau-induced neurodegeneration also deny causal relationship between NFTs formation and neuronal death. In a tau transgenic mouse in which the overexpression of mutant human tau can be regulated by tetracycline, turning off tau expression halts neuronal loss and reverses memory defects. But surprisingly, in this model, NFTs continue to accumulate, suggesting that NFTs are not responsible for neurodegeneration [18]. This result is consistent with reports on transgenic mice expressing nonmutant human tau as well as transgenic *Drosophila* expressing wild-type or mutant form of human tau, in which neurodegeneration or neuronal death occurs independently of NFT formation [19, 20].

4 Oxidative Stress Precedes Tau Pathology

Cellular [21, 22] and animal models [23–27] (Table 1) as well as human studies [30–36] (Table 2) suggest that oxidative stress chronologically precedes NFT formation. Oxidative stress activates several kinases including glycogen synthase kinase-3β (GSK-3β) and mitogen-activated protein kinases (MAPKs), which are activated in AD and are capable of phosphorylating tau. Once phosphorylated, tau becomes particularly vulnerable to oxidative modification and consequently aggregates into fibrils [37]. Therefore, NFT formation is likely to be a result of neuronal oxidation. Furthermore, in neurons of postmortem AD brains, a decrease in oxidative damage in nucleic acids (mainly cytoplasmic RNAs) is associated with the presence of NFTs, as determined by a comparison of neurons with and without NFTs, an observation that is particularly striking in light of the abundance of RNA on NFTs [33].

5 NFT Formation as a Compensatory Response
   to Oxidative Stress

One possible mechanism as to how NFT formation opposes oxidative stress may be associated with metal-binding capability of tau, in common with the capability of Aβ [15, 38]. Redox-active iron accumulation is strikingly associated with NFTs [39] and tau is found to be capable of binding to iron and copper and thereby possibly exerts antioxidant activities [40].

Additionally, tau and neurofilament proteins that are modified by lipid peroxidation products and carbonyls [41–43, 35] may work as a physiological "buffer" against toxic intermediates derived from oxidative reactions and thereby enhance neuronal survival. Although tau and neurofilaments are cytoskeletal proteins with long half-lives, the extent of carbonyl modification is comparable in young and aged mice, as well as along the length of the axon [44]. A logical explanation for this finding is that the oxidative modification of cytoskeletal proteins is under tight regulation. A high content of lysine-serine-proline (KSP) domains on both tau and neurofilament protein suggests that they are uniquely adapted to undergoing oxidative attack. Exposure of these domains on the protein surface is effected by extensive phosphorylation of the serine residues, resulting in an oxidative "sponge" of surface-accessible lysine residues, which are specifically modified by products of lipid peroxidation [44]. Because phosphorylation plays this pivotal role in redox balance, it is not surprising that oxidative stress leads to phosphorylation through activation of MAPK pathways [36, 45, 46], nor

Table 1 Evidence suggesting an involvement of oxidative stress in early steps of NFTs formation or tau-associated neurodegeneration in experimental models

Peptides, cellular, and animal models	Findings
Human tau prepared from autopsied brain tissue	Treatment of normal tau with HNE, a carbonyl product resulting from lipid peroxidation, significantly enhances the recognition of phosphoralation-dependent NFT antibodies and conformation-dependent antibodies, only when tau is in the phosphorylated state [28].
Neural cell culture	Treatment of cell cultures with acrolein, a lipid peroxidation product, significantly increases tau phosphorylation due to p38 stress-activated kinase [21].
Primary rat cortical neuron culture	Treatment of neuron cultures with cuprizone, a copper chelator, in combination with Fe^{2+}/H_2O_2 significantly increases tau phosphorylation with increased GSK activity [22].
Tau transgenic *Drosophila* (tau[R406W] heterozygous for SOD2 or for Trxr	Downregulation of SOD2 or Trxr antioxidant activities increases neuronal cell death in the model of human tauopathy, where tau phosphorylation is *not* promoted but tau-induced cell cycle activation is enhanced. The extent of JNK activation correlates with the degree of tau-induced degeneration [29].
Human tau transgenic mice	Vitamin E supplementation suppresses the development of tau pathology (filamentous tau aggregates) [26].
Triple-transgenic mice (AβPP[Swe], presenilin-1[M146V], and tau[P301L])	Levels of Aβ 1–40 and Aβ 1–42 as well as phosphorylated tau in the hippocampus are decreased by caloric restriction, which may be associated with a reduction of oxidative stress [24].
SOD2 null mice	SOD2 null mice die within the first week but survive by treatment with a catalytic antioxidant. The low-dose antioxidant-treated SOD2 null mice show striking elevations in the level of tau phosphorylation (at Ser-396) [25].
AβPP transgenic mice (Tg2576) heterozygous for SOD2	SOD2 deficiency exacerbates amyloid burden and results in synergistic increase in the levels of phospho-tau [25].
ApoE null mice	Dietary deprivation of folate and vitamin E (antioxidants), coupled with iron (pro-oxidant), fosters an increase in nonphospho- and phospho-tau within brain tissue [23].
Segmental trisomy 16 mouse model for Down syndrome (Ts1Cje), which has a subset of triplicated human chromosome 21 gene orthologues that exclude AβPP and SOD1	ROS generation and mitochondrial dysfunction are seen in neuron and astrocyte primary cultures derived from fetal Ts1Cje hippocampus. Tau hyperphosphorylation in brain starts at 2–3 months stage in Ts1Cje without NFT formation. GSK3β and JNK/SAPK are activated in Ts1Cje [27].

AβPP amyloid-β protein precursor, *ApoE* apolipoprotein E, *GSK3β* glycogen synthase kinase-3β, *HNE* 4-hydroxy-2-nonenal, *NFTs* neurofibrillary tangles, *ROS* reactive oxygen species, *SOD* superoxide dismutase, *Trxr* thioredoxin peroxidases

Table 2 Evidence suggesting an involvement of oxidative stress in early steps of NFTs formation in human brains

Human samples	Findings
Postmortem brains from subjects with Down syndrome	In a series of aging brains of Down syndrome cases, oxidative damages to neuronal RNA (8-OHG) and protein (3-NT) are prominent in the teens and twenties, which occurs prior to the formation of mature senile plaques and NFTs and increases in Aβ [34].
Postmortem brains from subjects with AD	1. Widespread oxidative damage to RNA (8-OHG) is detected in vulnerable neurons in AD. Indeed, the oxidative RNA damage is more prominent in neurons free of NFTs compared to neurons with NFTs, which is independent of cellular abundance of RNA [33].
	2. AGEs are always detected in neurons with diffuse, nonfibrillar hyperphosphorylated tau (positive for the AT-8 antibody), that is, pre-NFTs, whereas extraneuronal NFTs (end-stage NFTs) very rarely show AGEs [31].
	3. Intraneuronal appearance of cellular stress signals induced by oxidative and mitogenic stress such as activated ERK, JNK/SAPK, and p38 shows chronological and spatial relationship with progression of NFTs formation (Braak staging). In nondemented cases lacking pathology (Braak stage 0), either ERK alone or JNK/SAPK alone can be activated. In nondemented cases with limited pathology (Braak stages I and II), both ERK and JNK/SAPK are activated but p38 is not, while all three kinases are activated in the vulnerable neurons in mild and severe AD cases (Braak stages III-VI) [36].
	4. Distribution of an antioxidant enzyme HO-1-containing neurons shows a complete overlap with conformational change of tau (positive for the Alz50 antibody), but tau phosphorylation (positive for the AT8 antibody) occurs not only in these neurons but also in neurons not displaying HO-1, suggesting that the antioxidant HO-1 response follows tau phosphorylation and that HO-1 is coincident with the conformational change of tau [35].
Postmortem brains from subjects with MCI	Oxidative damage to RNA (8-OHG, NPrG) is always detected in neurons with conformational change of tau (positive for the MC-1 antibody), whereas oxidative RNA damage is detected also in minimal or no MC-1 immunostaining, suggesting RNA oxidation occurs prior to changes in tau conformation [30].
Postmortem brain from a presymptomatic case with presenilin-1 gene mutation	Oxidative damage to neuronal RNA (8-OHG) is increased in the frontal cortex where no neocortical NFTs are seen [32].

AGEs advanced glycation end-products, *ERK* extracellular receptor kinase, *HO-1* heme oxygenase-1, *MCI* mild cognitive impairment, *NFTs* neurofibrillary tangles, *NPrG* 1-*N*2-propanodeoxyguanosine, *3-NT* 3-nitrotyrosine, *8-OHG* 8-hydroxyguanosine

that conditions associated with chronic oxidant stress, such as AD, are associated with extensive phosphorylation of cytoskeletal elements.

Indeed, other tauopathies such as progressive supranuclear palsy, corticobasal degeneration, and frontotemporal dementia also show evidence of oxidative adducts on these proteins [47, 48]. This protective role of tau phosphorylation explains the finding that embryonic neurons that survive after treatment with oxidants have more phospho-tau immunoreactivity relative to neurons under degeneration [49]. Further, the induction of heme oxygenase, an antioxidant enzyme (which cleaves the oxidant heme), reduces tau expression and phosphorylation, indicating a crucial role for tau in redox homeostasis [50, 35]. Supporting this notion, there is reduced oxidative damage in neurons with tau accumulation that we suspect is due to the antioxidant function of phosphorylated tau.

6 Pathological Hallmarks and Their Neuroprotective Function: Aggregation-State Dependent?

"Aberrantly" (sic) folded proteins are common to a great number of neurodegenerative diseases and are, for the most part, vilified. The focus of Parkinson's disease, Pick disease, and amyotrophic lateral sclerosis, for example, has been on Lewy bodies, Pick bodies, and spheroids, their respective protein components. However, the concept that such intracellular inclusions are manifestations of cell survival may be a common feature of all neurodegenerative diseases. Such a notion, while heretical to most, recently found support in a Huntington's disease model [51]. In this neuronal model, cell death was mutant-huntingtin-dose- and polyglutamine-dependent; however, huntingtin inclusion formation correlated with cell *survival.* Thus, in this model, as in AD, inclusion formation represents adaptation, or a productive, beneficial response to the otherwise neurodegenerative process. Taken together with our studies, this represents a fundamental and necessary change in which pathological manifestations of neurodegenerative disease are interpreted.

As we have reviewed recently [38], disease-specific proteins such as Aβ, tau, and α-synuclein potentially play a protective role against oxidative stress. However, the efficiency of the protective function may be dependent on the concentrations or the aggregation state of the protein [52–55]. Recently, an increasing body of evidence has been collected to support the hypothesis that oligomers, not monomers or fibrils, represent the toxic form of Aβ, tau, and α-synuclein [56–60]. More detective work is required before this small intermediate fraction (oligomers) can be convicted as the real culprit. However, if only the oligomeric fraction is detrimental and monomeric peptides per se as well as mature fibrils are protective, therapeutic approaches targeting the protein should be highly specific for the oligomeric aggregation state. Therefore, further study is required to adequately assess the relationship between oxidative stress and oligomer formation, which may provide an important clue to early therapeutic intervention in neurodegenerative disorders.

7   Antioxidative Strategy for Neurodegenerative Disorders

Despite the abundant evidence for an involvement of oxidative insults as an early stage of the neurodegenerative process, interventions such as the administration of one or a few antioxidants have been, at best, modestly successful in clinical trials. The complexity of the metabolism of ROS suggests that such interventions may be too simplistic and requires more integrated approaches not only to enrich the exogenous antioxidants but also to upregulate the multilayered endogenous antioxidative defense systems [61, 15]. Recently expanding knowledge of the molecular mechanisms of organism longevity indicate that prolongevity gene products such as forkhead transcription factors and sirtuins are involved in the insulin-like signaling pathway and oxidative stress resistance against aging. An enhancement of the prolongevity signaling, which is possibly realized by caloric restriction or caloric restriction mimetics [62, 63], may be a promising approach in antioxidative strategy against age-associated neurodegenerative diseases.

In this context, a possible protective role of tau phosphorylation is particularly interesting because tau phosphorylation is induced by impaired insulin-like signaling and downstream activation of GSK-3β [64, 65]. Indeed, defect in insulin-like signaling is beneficial to longevity in diverse organisms at least partly through activation of endogenous antioxidant systems [66, 67]. Together with findings that reversible tau phosphorylation is an adaptive process associated with neuronal plasticity in hibernating animals [68], the involvement of tau phosphorylation in the insulin-like signaling pathway implicates an essential link between NFT formation and adaptation under oxidative stress in age-associated neurodegeneration.

8   Conclusions

In contrast to the general aspects of the pathological hallmarks, aggregation of the disease-specific protein in neurodegenerative disorders may be involved in a cellular compensatory response against oxidative insult. A recently increasing body of evidence suggests that tau aggregation and NFT formation are not exceptions for this new understanding of the classical pathologies and such notion may open novel therapeutic avenue to early intervention for AD and other tauopathies.

References

1. Hyman BT, Trojanowski JQ (1997) Consensus recommendations for the postmortem diagnosis of Alzheimer disease from the National Institute on Aging and the Reagan Institute Working Group on Diagnostic Criteria for the Neuropathological Assessment of Alzheimer Disease. J Neuropath Exp Neurol 56:1095–1097

2. Mirra SS, Heyman A, McKeel D et al. (1991) The Consortium to Establish a Registry for Alzheimer's Disease (CERAD) Part II. Standardization of the neuropathologic assessment of Alzheimer's disease. Neurology 41:479–486
3. Thal DR, Del Tredici K, Braak H. (2004) Neurodegeneration in normal brain aging and disease. Sci Aging Knowledge Environ 23:26
4. Davis DG, Schmitt FA, Wekstein DR, Markesbery WR. (1999) Alzheimer neuropathologic alterations in aged cognitively normal subjects. J Neuropath Exp Neurol 58:376–378
5. Arriagada PV, Growdon JH, Hedley-Whyte ET, Hyman BT. (1992) Neurofibrillary tangles but not senile plaques parallel duration and severity of Alzheimer's disease. Neurol 42:631–639
6. Gómez-Isla T, Hollister R, West H et al. (1997) Neuronal loss correlates with but exceeds neurofibrillary tangles in Alzheimer's disease. Ann Neurol 41:17–24
7. Neve RL, Robakis NK. (1998) Alzheimer's disease: a re-examination of the amyloid hypothesis. Trends Neurosci 21:15–19
8. Kril JJ, Patel S, Harding AJ, Halliday GM. (2002) Neuron loss from the hippocampus of Alzheimer's disease exceeds extracellular neurofibrillary tangle formation. Acta Neuropathol (Berl) 103:370–376
9. Cash AD, Aliev G, Siedlak SL et al. (2003) Microtubule reduction in Alzheimer's disease and aging is independent of tau filament formation. Am J Pathol 162:1623–1627
10. Alonso Adel C, Grundke-Iqbal I, Iqbal K. (1996) Alzheimer's disease hyperphosphorylated tau sequesters normal tau into tangles of filaments and disassembles microtubules. Nat Med 2:783–787
11. Alonso Adel C, Mederlyova A, Novak M, Grundke-Iqbal I, Ikbal K. (2004) Promotion of hyperphosphorylation by frontotemporal dementia tau mutations. J Biol Chem 279:34873–34881
12. Iqbal K, Alonso Adel C, Chen S et al. (2005) Tau pathology in Alzheimer disease and other tauopathies. Biochim Biophys Acta 1739:198–210
13. Morsch R, Simon W, Coleman PD. (1999) Neurons may live for decades with neurofibrillary tangles. J Neuropathol Exp Neurol 58:188–197
14. Ihara Y. (2001) PHF and PHF-like fibrils – cause or consequence? Neurobiol Aging 22:123–126
15. Nunomura A, Castellani RJ, Zhu X, Moreira PI, Perry G, Smith MA. (2006) Involvement of oxidative stress in Alzheimer's disease. J Neuropathol and Exp Neurol 65:631–641
16. Goedert M, Jakes R. (2005) Mutations causing neurodegenerative tauopathies. Biochim Biophys Acta 1739:240–250
17. Shiarli AM, Jennings R, Shi J et al. (2006) Comparison of extent of tau pathology in patients with frontotemporal dementia with Parkinsonism linked to chromosome 17 (FTDP-17), frontotemporal lobar degeneration with Pick bodies and early onset Alzheimer's disease. Neuropathol Appl Neurobiol 32:374–387
18. Santacruz K, Lewis J, Spires T et al. (2005) Tau suppression in a neurodegenerative mouse model improves memory function. Science 309:476–481
19. Andorfer C, Acker CM, Kress Y, Hof PR, Duff K, Davies P. (2005) Cell-cycle reentry and cell death in transgenic mice expressing nonmutant human tau isoforms. J Neurosci 25:5446–5454
20. Wittmann CW, Wszolek MF, Shulman JM et al. (2001) Tauopathy in Drosophila: neurodegeneration without neurofibrillary tangles. Science 293:711–714
21. Gomez-Ramos A, Diaz-Nido J, Smith MA, Perry G, Avila J. (2003). Effect of the lipid peroxidation product acrolein on tau phosphorylation in neural cells. J Neurosci Res 71:863–870
22. Lovell MA, Xiong S, Xie C, Davies P, Markesbery WR. (2004) Induction of hyperphosphorylated tau in primary rat cortical neuron cultures mediated by oxidative stress and glycogen synthase kinase-3. J Alzheimers Dis 6:659–671
23. Chan A, Shea TB. (2006) Dietary and genetically-induced oxidative stress alter tau phosphorylation: influence of folate and apolipoprotein E deficiency. J Alzheimers Dis 9:399–405
24. Halagappa VK, Guo Z, Pearson M et al. (2007) Intermittent fasting and caloric restriction ameliorate age-related behavioral deficits in the triple-transgenic mouse model of Alzheimer's disease. Neurobiol Dis 26:212–220
25. Melov S, Adlard PA, Morten K et al. (2007) Mitochondrial oxidative stress causes hyperphosphorylation of tau. PLoS ONE 2(6):e536

26. Nakashima H, Ishihara T, Yokota O et al. (2004) Effects of alpha-tocopherol on an animal model of tauopathies. Free Radic Biol Med 37:176–186
27. Shukkur EA, Shimohata A, Akagi T et al. (2006) Mitochondrial dysfunction and tau hyperphosphorylation in Ts1Cje, a mouse model for Down syndrome. Hum Mol Genet 15:2752–2762
28. Liu Q, Smith MA, Avila J et al. (2005) Alzheimer-specific epitopes of tau represent lipid peroxidation-induced conformations. Free Radic Biol and Medic 38:746–754
29. Dias-Santagata D, Fulga TA, Duttaroy A, Feany MB. (2007) Oxidative stress mediates tau-induced neurodegeneration in Drosophila. J Clin Invest 117(1):236–245
30. Lovell MA, Markesbery WR. (2008) Oxidatively modified RNA in mild cognitive impairment. Neurobiol Dis 29:169–175
31. Lüth HJ, Ogunlade V, Kuhla B et al. (2005) Age- and stage-dependent accumulation of advanced glycation end products in intracellular deposits in normal and Alzheimer's disease brains. Cereb Cortex 15:211–220
32. Nunomura A, Chiba S, Lippa CF et al. (2004) Neuronal RNA oxidation is a prominent feature of familial Alzheimer's disease. Neurobiol of Dis 17:108–113
33. Nunomura A, Perry G, Aliev G et al. (2001) Oxidative damage is the earliest event in Alzheimer disease. J Neuropathol Exp Neurol 60:759–767
34. Nunomura A, Perry G, Pappolla MA et al. (2000) Neuronal oxidative stress precedes amyloid-beta deposition in Down syndrome. J Neuropathol Exp Neurol 59:1011–1017
35. Takeda A, Smith MA, Avila J et al. (2000) In Alzheimer's disease, heme oxygenase is coincident with Alz50, an epitope of tau induced by 4-hydroxy-2-nonenal modification. J Neurochem 75:1234–1241
36. Zhu X, Castellani RJ, Takeda A et al. (2001) Differential activation of neuronal ERK, JNK/SAPK and p38 in Alzheimer disease: the 'two hit' hypothesis. Mech Ageing Dev 123:39–46
37. Lee HG, Perry G, Moreira PI et al. (2005) Tau phosphorylation in Alzheimer's disease: pathogen or protector? Trends Mol Med 11:164–169
38. Nunomura A, Moreira PI, Lee HG et al. (2007) Neuronal death and survival under oxidative stress in Alzheimer and Parkinson diseases. CNS Neurol Disord Drug Targets 6:411–423
39. Smith MA, Harris PL, Sayre LM, Perry G. (1997) Iron accumulation in Alzheimer disease is a source of redox-generated free radicals. Proc Natl Acad Sci USA 94:9866–9868
40. Sayre LM, Perry G, Harris PL, Liu Y, Schubert KA, Smith MA. (2000) In situ oxidative catalysis by neurofibrillary tangles and senile plaques in Alzheimer's disease: a central role for bound transition metals. J Neurochem 74:270–279
41. Calingasan NY, Uchida K, Gibson GE. (1999) Protein-bound acrolein: a novel marker of oxidative stress in Alzheimer's disease. J Neurochem 72: 751–756
42. Sayre LM, Zelasko DA, Harris PL, Perry G, Salomon RG, Smith MA. (1997) 4-Hydroxynonenal-derived advanced lipid peroxidation end products are increased in Alzheimer's disease. J Neurochem 68:2092–2097
43. Smith MA, Rudnicka-Nawrot M, Richey PL et al. (1995) Carbonyl-related posttranslational modification of neurofilament protein in the neurofibrillary pathology of Alzheimer's disease. J Neurochem 64:2660–2666
44. Wataya T, Nunomura A, Smith MA et al. (2002) High molecular weight neurofilament proteins are physiological substrates of adduction by the lipid peroxidation product hydroxynonenal. J Biol Chem 277:4644–4648
45. Zhu X, Raina AK, Rottkamp CA et al. (2001) Activation and redistribution of c-jun N-terminal kinase/stress activated protein kinase in degenerating neurons in Alzheimer's disease. J Neurochem 76:435–441
46. Zhu X, Rottkamp CA, Boux H, Takeda A, Perry G, Smith MA (2000) Activation of p38 kinase links tau phosphorylation, oxidative stress, and cell cycle-related events in Alzheimer disease. J Neuropathol Exp Neurol 59:880–888
47. Castellani R, Smith MA, Richey PL, Kalaria R, Gambetti P, Perry G. (1995) Evidence for oxidative stress in Pick disease and corticobasal degeneration. Brain Res 696:268–271
48. Odetti P, Garibaldi S, Norese R et al. (2000) Lipoperoxidation is selectively involved in progressive supranuclear palsy. J Neuropathol Exp Neurol 59:393–397

49. Ekinci FJ, Shea TB. (2000) β-amyloid-induced tau phosphorylation does not correlate with degeneration in cultured neurons. J Alzheimers D*is* 2:7–15
50. Takeda A, Perry G, Abraham NG et al. (2000) Overexpression of heme oxygenase in neuronal cells, the possible interaction with Tau. J Biol Chem 275:5395–5399
51. Arrasate M, Mitra S, Schweitzer ES, Segal MR, Finkbeiner S. (2004) Inclusion body formation reduces levels of mutant huntingtin and the risk of neuronal death. Nature 431:805–810
52. Albani D, Peverelli E, Rametta R et al. (2004) Protective effect of TAT-delivered alpha-synuclein: relevance of the C-terminal domain and involvement of HSP70. FASEB J 18:1713–1715
53. Kontush, A. (2001) Amyloid-β: an antioxidant that becomes a pro-oxidant and critically contributes to Alzheimer's disease. Free Radic Biol Med 31:1120–1131
54. Seo JH, Rah JC, Choi SH et al. (2002) Alpha-synuclein regulates neuronal survival via Bcl-2 family expression and PI3/Akt kinase pathway. FASEB J 16:1826–1828
55. Zou K, Gong JS, Yanagisawa K, Michikawa M. (2002) A novel function of monomeric amyloid β-protein serving as an antioxidant molecule against metal-induced oxidative damage. J Neurosci 22:4833–4841
56. Maeda S, Sahara N, Saito Y, Murayama S, Ikai A, Takashima A. (2006) Increased levels of granular tau oligomers: an early sign of brain aging and Alzheimer's disease. Neurosci Res 54:197–201
57. McLean CA, Cherny RA, Fraser FW et al. (1999) Soluble pool of Abeta amyloid as a determinant of severity of neurodegeneration in Alzheimer's disease. Ann Neurol 46:860–866
58. Sato S, Tatebayashi Y, Akagi T et al. (2002) Aberrant tau phosphorylation by glycogen synthase kinase-3β and JNK3 induces oligomeric tau fibrils in COS-7 cells. J Biol Chem 277:42060–42065
59. Sharon R, Bar-Joseph I, Frosch MP, Walsh DM, Hamilton JA, Selkoe DJ. (2003) The formation of highly soluble oligomers of alpha-synuclein is regulated by fatty acids and enhanced in Parkinson's disease. Neuron 37:583–595
60. Walsh DM, Klyubin I, Fadeeva JV et al. (2002) Naturally secreted oligomers of amyloid beta protein potently inhibit hippocampal long-term potentiation in vivo. Nature 416:535–539
61. Lin MT, Beal MF. (2006) Mitochondrial dysfunction and oxidative stress in neurodegenerative diseases. Nature 443:787–795
62. Baur JA, Pearson KJ, Price NL et al. (2006) Resveratrol improves health and survival of mice on a high-calorie diet. Nature 444:337–342
63. Heilbronn LK, de Jonge L, Frisard MI et al. (2006) Effect of 6-month calorie restriction on biomarkers of longevity, metabolic adaptation, and oxidative stress in overweight individuals: a randomized controlled trial. JAMA 295:1539–1548
64. Cheng CM, Tseng V, Wang J, Wang D, Matyakhina L, Bondy CA. (2005) Tau is hyperphosphorylated in the insulin-like growth factor-I null brain. Endocrinol 146:5086–5091
65. Hooper C, Killick R, Lovestone S. (2008) The GSK3 hypothesis of Alzheimer's disease. J Neurochem 104:1433–1439
66. Finkel T, Holbrook NJ. (2000) Oxidants, oxidative stress and the biology of ageing. Nature 408:239–247
67. Tatar M, Bartke A, Antebi A. (2003) The endocrine regulation of aging by insulin-like signals. Science 299:1346–1351
68. Arendt T, Stieler J, Strijkstra AM et al. (2003) Reversible paired helical filament-like phosphorylation of tau is an adaptive process associated with neuronal plasticity in hibernating animals. J Neurosci 23:6972–6981

Neuroimmunological Hypothesis

Inflammatory Processes Exacerbate Degenerative Neurological Disorders

Patrick L. McGeer, Edith G. McGeer, and Claudia Schwab

Abstract Evidence that neuroinflammation exacerbates the pathology in Alzheimer's disease (AD) has been accumulated from three independent fields of research: neuropathology demonstrating activated glial cells, epidemiology showing sparing of AD in individuals consuming anti-inflammatory drugs, and AD transgenic animal studies showing protection from COX-1-inhibiting drugs. Similar findings have now been established for Parkinson's disease. Inflammation accompanies the lesions; epidemiological studies show a sparing effect of NSAIDs; and anti-inflammatory agents are protective in animal models of the disease. Inflammation is not believed to be the triggering event in these conditions, but nevertheless is capable of chronically exacerbating the pathology. It is characterized by the generation of a spectrum of inflammatory mediators produced locally by resident cells. Such local production indicates engagement of the innate immune system. Similar phenomena are now being observed in degenerative peripheral conditions such as diabetes type 2. A distinction needs to be made between these autotoxic disorders and classical autoimmune disorders, which involve participation of the adaptive immune system. The adaptive immune system is more powerful, resulting in a range of autoimmune diseases that are more severe and strike at a younger age. Autotoxic diseases, being milder, occur in the aging population but overall are more prevalent than autoimmune diseases. Both immune systems use local phagocytes such as microglia to be the effecter cells so that methods to reduce the toxic effects of their overstimulation should be beneficial in a broad spectrum of human diseases.

Keywords Microglia, Autotoxicity, Autoimmunity, Innate immune system, Alzheimer's disease, Age-related macular degeneration, Diabetes, Atheroscelorsis, Parkinson's disease

P.L. McGeer (✉), E.G. McGeer, and C. Schwab
Kinsmen Laboratory of Neurological Research
University of British Columbia
2255 Wesbrook Mall,
Vancouver, BC V6T1Z3, Canada

R.B. Maccioni and G. Perry (eds.) *Current Hypotheses and Research Milestones in Alzheimer's Disease*. DOI: 10.1007/978-0-387-87995-6_10,
© Springer Science + Business Media, LLC 2009

117

All inflammatory reactions are also immune reactions. It is the innate immune system that is first called into action. Later, the adaptive system may also respond. Chronic inflammation signifies that an immune reaction is being sustained. That is because healing has failed to take place. Chronic inflammation may involve the innate immune system, the adaptive immune system, or a combination of the two.

Each disease has its own etiology, but the body's defensive reactions to them involve the same immune mechanisms. The final effecter cells for both the innate and adaptive immune systems are tissue-based phagocytes. When chronic inflammatory reactions are relatively low grade and localized, it is the innate system that is primarily involved. When the reactions are more powerful and systemic, it is the adaptive system that is primarily involved. In either case, diseases causing chronic inflammatory responses tend to be progressive. Without definitive methods of treatment, they are usually fatal.

An important contributor to disease progression is self-damage inflicted by over activity of effector cells. In attempting to rid the body of disease-causing factors, they are capable of injuring viable host tissue. It is this phenomenon which forms the basis of anti-inflammatory treatment of many chronic diseases. Phagocytes elaborate a spectrum of inflammatory stimulants that are capable of damaging viable tissue. These toxic factors include free radicals, complement components, inflammatory cytokines, prostaglandins, excess glutamate, various proteases, and a spectrum of other molecules known as inflammatory mediators.

Chronic degenerative diseases involving self-damage fall into two broad categories: autoimmune and autotoxic [1]. Autoimmune diseases are classically defined as those where evidence exists of involvement of the adaptive immune system. The involvement may be humoral, in which B-lymphocytes are cloned to produce and secrete antibodies against self-proteins, or it may be cell mediated, in which T cells are cloned to generate an attack against cells expressing identified protein epitopes. Autotoxic diseases are a more recently defined category [2]. Self-attack also occurs, but not because of engagement of the adaptive immune system. It occurs because the innate system, acting locally, is generating the attack. Autotoxic diseases typically attack the elderly rather than the young to middle-aged people who are prone to autoimmune diseases. That is because the innate immune system is less powerful and less focused than the adaptive immune system. Nevertheless, in sum, autotoxic degenerative disorders are more prevalent than autoimmune disorders.

Table 1 Examples of autoimmune and autotoxic disorders

Some autoimmune diseases	Some autotoxic diseases
Systemic lupus	Tauopathies
Myasthenia gravis	Alzheimer's disease
Rheumatoid arthritis	Parkinson's disease
Diabetes type 1	Atherosclerosis
Multiple sclerosis	Amyotrophic lateral sclerosis
Lambert Eaton syndrome	Diabetes type 2
	Macular degeneration

Table 1 lists some typical examples of autoimmune and autotoxic disorders. Typical autoimmune diseases include systemic lupus erythematosis, myasthenia gravis, rheumatoid arthritis, diabetes type 1, multiple sclerosis, and the Lambert Eaton syndrome. Typical autotoxic disorders include Alzheimer's disease (AD), Parkinson's disease (PD), age-related macular degeneration (AMD), diabetes type 2, atherosclerosis, amyotrophic lateral sclerosis (ALS), and the tauopathies.

AD is the prototypical autotoxic disorder. Classical immunologists had originally declared that it was a sterile, noninflammatory degenerative condition. The conclusion was based on the absence of infiltrating lymphocytes and monocytes which were easily observed in CNS infections and in presumed autoimmune diseases such as multiple sclerosis. This concept was challenged by the identification of activated microglia expressing HLA-DR in association with the lesions of AD [3, 4]. HLA-DR had already been recognized as a marker of immunocompetent cells. This observation vindicated Hortega's original conclusion in 1919 that microglia were phagocytic cells of mesodermal origin. Moreover, it established that chronic inflammation could exist in the absence of leukocyte infiltration.

Two subsequent steps showed that the inflammation was self-damaging. The first was immunohistochemical, in which dystrophic neurites being damaged by the membrane attack complex of complement could be observed in AD tissue [5–7]. The second was epidemiological, in which those taking anti-inflammatory agents appeared to be spared from AD [8]. If the inflammation observed immunohisto-chemically had been beneficial instead of harmful, taking anti-inflammatories should have increased the risk of AD, and if they were merely cleaning up debris, then they should have had no effect. These epidemiological findings, which have been replicated in more than 20 studies, clearly show that the inflammation is con-tributing to the disease pathology, and that long-term consumption of NSAIDs reduces the risk of AD from two- to fivefold [9, 10].

Further confirmation comes from studies in AD transgenic mouse models where the administration of classical NSAIDs reduces the behavioral deficits as well as the lesion burden caused by overexpression of amyloid-β (Aβ) [11].

Any reservations that might still be held that microglia are the sentinels of the brain that respond to disease pathology should have been dispelled by the recent in vivo movies of Nimmerjahn et al. [12]. They labeled mouse microglia with green fluorescent protein and then carried out time-lapse photography of their activity through the exposed surface of the brain. They observed that resting microglia were never at rest. They were in constant activity, testing for abnormalities by continu-ously sampling their surround. When an abnormality was induced by laser damage to a capillary, the microglia immediately changed to an activated morphology. Within minutes, they began migrating to the lesion site where they sealed off the affected area and commenced phagocytosing the extravasated blood.

Microglia are the brain representatives of van Furth's monocyte phagocytic system [13]. Their counterparts, such as Langerhan's cells of the skin, Kupfer cells of the liver, osteoclasts of bone, and macrophages in many organs, are tissue-based phagocytes which can be presumed to carry out similar functions to microglia throughout the body. While it is only speculation at this stage, it might be anticipated that all cells of the monocyte phagocytic system would respond to pathological challenges in a similar fashion to those observed by Nimmerjahn et al. of brain microglia [12].

After the discovery of HLA-DR reactive microglial cells in AD, it was soon shown that such microglia were associated with the lesions of PD, Pick disease, Huntington disease, ALS, and the parkinsonism dementia complex of Guam, all of which had been considered to be noninflammatory disorders [14].

One prediction emerging from these data is that anti-inflammatories should reduce the risk of PD. PD is less common than AD so that epidemiological data have been more difficult to acquire. Nevertheless, there are now multiple studies showing a protective effect of NSAIDs [15, 16].

AMD is another common disease of the elderly. It is estimated to affect more than 10 million individuals in the USA alone [17]. The premonitory signs of AMD are the appearance of drusen, which are small, extracellular deposits that are somewhat similar to AD senile plaques. They develop at the junction between the choroid plexus and the retinal pigment epithelium [18, 19]. There is a well-documented inflammatory reaction accompanying drusen development. Inflammatory markers such as C-reactive protein (CRP), activated microglia, and activated complement fragments, including the membrane attack complex, have been identified in association with the lesions [18]. This has led to the hypothesis that the localized inflammatory reaction in AMD is analogous to that observed in AD and atherosclerosis, and that the inflammation exacerbates the pathogenesis.

On this basis, we identified a tenfold sparing of AMD in rheumatoid arthritics compared with controls [20]. Of more fundamental interest has been a series of reports showing that polymorphisms in factor H can have as much as a tenfold influence on the risk of AMD [21]. Factor H is an inhibitory factor for the alternative complement pathway, acting to protect the host from self-damage caused by excessive activation of the alternative complement pathway. There is one significant difference from AD. In AMD, it is the alternative pathway which is activated, while in AD, Aβ causes excess activation of the classical pathway [6].

Type 2 diabetes is another common disease of late middle age where inflammation appears to play a significant role. It is the most common form of diabetes, comprising 90–95% of all diabetic cases [22]. It is characterized by a slowly progressive degeneration of islet β-cells, resulting in a fall of insulin secretion. The pancreas accumulates deposits of amylin, which is a 37-amino acid peptide derived by proteolytic cleavage of islet amyloid precursor protein. The amylin deposits are fibrillar in nature and there is a similarity in structure to Aβ [23]. There is also an overlap with AD. The percentage of type 2 diabetes cases among AD patients is significantly higher than among age-matched non-AD controls [24]. Conversely, patients with type 2 diabetes have twice the risk of controls to develop AD [25]. The resident phagocytes of the pancreas become activated in type 2 diabetes, in similar fashion to brain microglia in AD and other neurodegenerative disorders. In vitro, the classical complement pathway is activated by amylin just as it is by Aβ [26].

Atherosclerosis may be the most common degenerative disease of aging. As in the previous conditions described, it is characterized by inflammation, in this case in the atheromatous lesions. Complement is activated, presumably by CRP which is an in vitro complement activator. CRP is upregulated in arterial cells in the lesioned area [27]. Resident macrophages become activated, and in that state, secrete collagenase type IV. This is particularly dangerous because it can dissolve the thin fibrin coat which covers the lesion. When this occurs, the inflamed contents are released, precipitating a thrombosis.

Figure 1 illustrates nonactivated and activated resident phagocytes in various disease situations immunostained by the CR3/43 monoclonal antibody against HLA-DR. Figure 1a shows faintly stained normal brain microglia in the globus pallidus, an unaffected area of an AD case. In contrast, Fig. 1b shows activated microglia in the heavily affected entorhinal cortex of the same case. Figure 1c shows no staining in an unaffected area of the aorta in an elderly case. Figure 1d shows activated macrophages in a nearby atherosclerotic plaque. Figure 1e shows that no HLA-DR staining is detectable in the unaffected pancreas of an elderly case, while Fig. 1f shows activated phagocytes in a nearby inflamed area.

Collectively, these data illustrate the widespread nature of autotoxic pathology and the enormous potential of anti-inflammatory therapy. To date, NSAIDs are by far the most widely utilized class of anti-inflammatory agents. They inhibit prostaglandin production, as do the steroids, but the prostaglandins are weak inflammatory mediators so this is a relatively ineffective target. Side effects of these agents are also a significant problem since prostaglandins and steroids both have important functions in normal physiology.

Inflammatory cytokines are more powerful mediators and are therefore a better target. Enbrel and Remicade, which block TNF, have had great success in treating rheumatoid arthritis. They are globulins which do not cross the blood–brain barrier and are therefore not suitable for CNS diseases. Nevertheless, they illustrate the promise of small molecules which could reach the brain and block the actions of TNF. Presumably, blockers of IL-1 and IL-6 would also be effective. Blockers of complement activation, especially the membrane attack complex, are also attractive targets for future therapy. As well, blockers of intracellular pathways that promote DNA transcription of inflammatory mediators should be effective. There are numerous reviews that cover the toxic effects of inflammatory mediators and the spectrum of possible agents or pathways to block in order to reduce such effects [11, 28, 29].

Attention is now being focused on a previously overlooked mode of microglial activation. That is the anti-inflammatory mode where downregulating cytokines such as IL-4 and IL-10 are released. This is thought to be in analogy to the Th-1 and Th-2 states of T cells and presumably reflects a switch from an attack state to a healing state. It is thought that phagocytosis may be enhanced in this state. Accordingly, there may be therapeutic scope in defining the intracellular pathways involved and exploring for stimulants related to those pathways. The possibilities have been the subject of a recent review by Schwab et al. [30].

In summary, exploration of the inflammatory aspects of AD pathology led first to the discovery that inflammation was associated with the lesions in a range of age-related neurodegenerative diseases and then to the discovery that it applied to

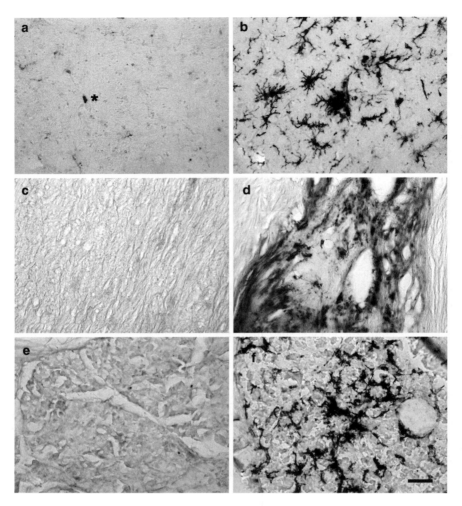

Fig. 1 Microphotographs of resident phagocytes in various disease situations, stained immuno-histochemically with an antibody against HLA-DR (CR3/43, DAKO). **a** shows very weak staining of normal brain microglia and one strongly stained pericyte (asterisks) in the unaffected globus pallidus of an AD case. In contrast, **b** shows many activated microglia aggregating around Aβ plaques in the entorhinal cortex of the same case. **c** shows no staining in an unaffected area of the aorta in an elderly case. **d** shows activated macrophages in a nearby atherosclerotic plaque. There is very little staining in unaffected pancreas (**e**) while **f** shows activated phagocytes in a nearby inflamed area. Calibration bar: 50 μm

age-related degenerative diseases of the periphery. Epidemiological studies, where a reduced risk of AD and PD was observed in those taking anti-inflammatory agents, established that this inflammation was exacerbating the fundamental pathology.

Innate but not adaptive immune forces have been identified as contributing to the toxic effects. Accordingly, these diseases have been described as autotoxic to distinguish them from classical autoimmune diseases where the adaptive immune system drives the pathology.

Our current understanding of the pathogenesis of autotoxic disorders should permit rational development of new and very powerful broad-spectrum therapeutic agents. Targets include inflammatory cytokines, complement factors, other inflammatory mediators, blockers of intracellular pathways that induce transcription of inflammatory mediators, and stimulation of intracellular pathways that induce transcription of anti-inflammatory mediators.

Acknowledgments This work was supported by a grant from the Pacific Alzheimer Research Foundation. The Drs. McGeer are named on patents held by the University of British Columbia for the treatment of dementia with cyclooxygenase inhibitors.

References

1. McGeer PL, McGeer EG (2001) Inflammation, autotoxicity and Alzheimer disease. Neurobiol Aging 22:799–809
2. McGeer PL, McGeer EG (2000) Autotoxicity and Alzheimer disease. Arch Neurol 57:789–790
3. McGeer PL, Itagaki S, Tago H et al. (1987) Reactive microglia in patients with senile dementia of the Alzheimer type are positive for the histocompatibility glycoprotein HLA-DR. Neurosci Lett 79:195–200
4. Rogers J, Luber-Narod J, Styren SD et al. (1988) Expression of immune system-associated antigens by cells of the human central nervous system: relationship to the pathology of Alzheimer's disease. Neurobiol Aging 9:339–349
5. McGeer PL, Akiyama H, Itagaki S et al. (1989) Activation of the classical complement pathway in brain tissue of Alzheimer patients. Neurosci Lett 107:341–346
6. Rogers J, Cooper NR, Webster S et al. (1992) Complement activation by beta-amyloid in Alzheimer disease. Proc Natl Acad Sci USA 89:10016–10020
7. Webster S, Lue LF, Brachova L et al. (1997) Molecular and cellular characterization of the membrane attack complex, C5b-9, in Alzheimer's disease. Neurobiol Aging 18:415–421
8. McGeer PL, McGeer EG, Rogers J et al. (1990) Anti-inflammatory drugs and Alzheimer disease. Lancet 335:1037
9. In TV, Ruitenberg A, Hofman A et al. (2001) Nonsteroidal anti-inflammatory drugs and the risk of Alzheimer's disease. N Engl J Med 345:1515–1521
10. Stewart WF, Kawas C, Corrada M et al. (1997) Risk of Alzheimer's disease and duration of NSAID use. Neurology 48:626–632
11. McGeer PL, McGeer EG (2007) NSAIDs and Alzheimer disease: epidemiological, animal model and clinical studies. Neurobiol Aging 28:639–647
12. Nimmerjahn A, Kirchhoff F, Helmchen F (2005) Resting microglial cells are highly dynamic surveillants of brain parenchyma in vivo. Science 308:1314–1318
13. Van Furth R (1998) Human monocytes and cytokines. Res Immunol 149:719–720
14. McGeer PL, Itagaki S, McGeer EG (1988) Expression of the histocompatibility glycoprotein HLA-DR in neurological disease. Acta Neuropathol (Berl) 76:550–557
15. Chen H, Zhang M, Hernan MA et al. (2003) Nonsteroidal anti-inflammatory drugs and the risk of Parkinson disease. Arch Neurol 60:1059–1064

16. Esposito E, Di Matteo V, Benigno A et al. (2007) Non-steroidal anti-inflammatory drugs in Parkinson's disease. Exp Neurol 205:295–312
17. Gottlieb JL (2002) Age-related macular degeneration. JAMA 288:2233–2236
18. Anderson DH, Mullins RF, Hageman GS et al. (2002) A role for local inflammation in the formation of drusen in the aging eye. Am J Ophthalmol 134:411–431
19. Penfold PL, Madigan MC, Gillies MC et al. (2002) Immunological and aetiological aspects of macular degeneration. Prog Retin Eye Res 20:385–414
20. McGeer PL, Sibley J (2005) Sparing of age-related macular degeneration in rheumatoid arthritis. Neurobiol Aging 26:1199–1203
21. Hageman GS, Anderson DH, Johnson LV et al. (2005) A common haplotype in the complement regulatory gene factor H (HF1/CFH) predisposes individuals to age-related macular degeneration. Proc Natl Acad Sci USA 102:7227–7232
22. Biessels GJ, Staekenborg S, Brunner E et al. (2006) Risk of dementia in diabetes mellitus: a systematic review. Lancet Neurol 5:64–74
23. Dahabada HJL, Colin C, Degaki TL et al. (2004) Amyloidogenicity and cytotoxicity of recombinant mature human islet amyloid polypeptide (rhIAPP). J Biol Chem 279:42803–42810
24. Janson J, Laedtke T, Parisi JE et al. (2004) Increased risk of type 2 diabetes in Alzheimer disease. Diabetes 53:474–481
25. Arvanitakis Z, Wilson RS, Bienias JL (2004) Diabetes mellitus and risk of Alzheimer disease and decline in cognitive function. Arch Neurol 61:661–666
26. Klegeris A, McGeer PL (2007) Complement activation by islet amyloid polypeptide (IAPP) and alpha-synuclein 112. Biochemical and Biophysical Research Communications 357:1096–1099
27. Yasojima K, Schwab C, McGeer EG et al. (2001) Generation of C-reactive protein and complement components in atherosclerotic plaques. Am J Pathol 158:1039–1051
28. Akiyama H, Barger S, Barnum S et al. (2000) Inflammation and Alzheimer's disease. Neurobiol Aging 21:383–421
29. Eikelenboom P, Van Gool WA (2004) Neuroinflammatory perspectives on the two faces of Alzheimer's disease. J Neural Transm 111:281–294
30. Schwab C, McGeer PL (2008) Inflammatory aspects of Alzheimer disease and other neurodegenerative disorders. J Alzheimers Dis 13:359–369

Central Nervous System Inflammation and Cholesterol Metabolism Alterations in the Pathogenesis of Alzheimer's Disease and Their Diagnostic and Therapeutic Implications

Leonel E. Rojo, Jorge Fernández, José Jiménez, Andrea A. Maccioni, Alejandra Sekler, Rodrigo O. Kuljis, and Ricardo B. Maccioni

Abstract During the last few years, an increasing amount of evidence points to the major role of deregulation of the interaction patterns between glial cells and neurons in the pathway toward neuronal degeneration. Central nervous system inflammation is a process associated with several neurodegenerative disorders, including Alzheimer's disease (AD). Many hypotheses have been postulated to explain the pathogenesis of AD. Recent findings point to amyloid-β (Aβ) oligomers as responsible for synaptic impairment in neuronal degeneration, but amyloid abnormalities are among major factors affecting the function and survival of neuronal cells. The tau protein hypothesis has been developed and refined based on the fact that tau hyperphosphorylation and self-aggregation constitutes a common feature of most of the altered signaling pathways in AD. Known mediators of inflammation have been found in plaques, such as interleukin-1β, interleukin-6, and tumor necrosis factor-α. Additional evidence for the involvement of inflammation in AD is provided by epidemiological data and retrospective clinical data showing positive effects of nonsteroidal anti-inflammatory drugs. Cytokines and trophic factors produced by glial cells can trigger anomalous hyperphosphorylation of tau. Glial production of these mediators indicates that innate immunity is involved in AD. Thus, a neuroimmunological approach to AD becomes relevant. In this context, endogenous danger signals such as altered lipoproteins and oxidized lipids appear to affect glial cells, inducing release of such mediators. Indeed, when alterations of cholesterol metabolism occur, the neurochemical events of oxidative stress, Aβ peptide, and tau protein seem to represent a set of physiological mechanisms to respond to impaired brain cholesterol dynamics. All these mechanisms, for example, changes in neuroimmunomodulation, dislipidemias, cholesterol abnormalities, and other metabolic

L.E. Rojo (✉), J. Fernández, J. Jiménez, A.A. Maccioni, A. Sekler, R.O. Kuljis, and R.B. Maccioni
Laboratory of Cell and Molecular Neuroscience
DCPSc of Arturo Prat University and Mind/Brain Program of the ICC
Las Encinas 3370,
Ñuñoa, Santiago, Chile

R.B. Maccioni and G. Perry (eds.) *Current Hypotheses and Research Milestones in Alzheimer's Disease*. DOI: 10.1007/978-0-387-87995-6_11,
© Springer Science + Business Media, LLC 2009

alterations, appear to be interrelated. To date, there are no specific diagnostic tools for AD that allow early treatment, thus improving quality of life for AD patients and reducing the morbidity and mortality associated with the late complications. Therefore, a search for innovative molecular markers for early diagnosis of AD is essential. Here we discuss the molecular aspects of the role of neuroinflammation and cholesterol in AD and some perspectives toward molecular early diagnosis.

1 Introduction

Inflammation in the central nervous system (CNS) is a complex process, mainly affecting neuron–glial interactions in a manner that creates a new homeostasis, where the normal patterns of molecular signals and trophic factors delivered by glial cells change, thus activating pathways that conduct to neurodegeneration. Current experimental and clinical evidence suggest that the release of endogenous damage/alarm signals in response to converging and accumulated cell distress (dyslipidemia, vascular insults, head injury, oxidative stress, folate deficiency, etc.) is the earliest triggering event in Alzheimer's disease (AD) pathogenesis. This in turn leads to the activation of innate immunity and subsequently triggers an inflammatory cascade mediated by cytokines released by microglial cells [1]. In this context, the progression of AD from its earliest stages is characterized by a slowly increasing damage of brain parenchyma that precedes the onset of symptoms and the eventual irreversible pathological events of AD (Fig. 1).

In the first part of this chapter, we discuss the multiple roles – protective and deleterious – of proinflammatory cytokines in the neurodegeneration process, and the relationships between CNS and peripheral cytokines levels in order to provide a rational background for the interpretation of current clinical and epidemiological data. The information derived from clinical studies suggests that alteration of neuron–glial interactions during AD and other dementia mainly involves an increase in the levels of tumor necrosis factor-α (TNFα), interleukin-6 (IL-6), IL-2, nitric oxide synthase (NOS), and the soluble CD40 ligand (sCD40s), as early events during the neurodegenerative process [2]. Therefore, we critically review the evidence on peripheral levels of these cytokines in biological fluids and early diagnosis of AD. In this analysis, the strength of scientific evidence leads us to emphasize the role of IL-6, IL-1β, CD40s, and TNFα among the most promising biological markers for AD. Since the levels of these molecules in biological fluids, when analyzed as potential markers, vary dramatically among different reports, it is crucial that any reliable biomarker for early AD should be based on the association of more than one specific marker. Thus, the use of a combined biomarker, for example, the one based on tumoral growth factor-β (TGFβ1) and NOS measurements in leukocytes recently reported by Deservi [3], seems to provide a more reliable correlation on the course of progression of cognitive impairment. Thus, development of novel diagnostic tools based on complex correlations of the polymorphisms of these proinflammatory markers is an area of interest.

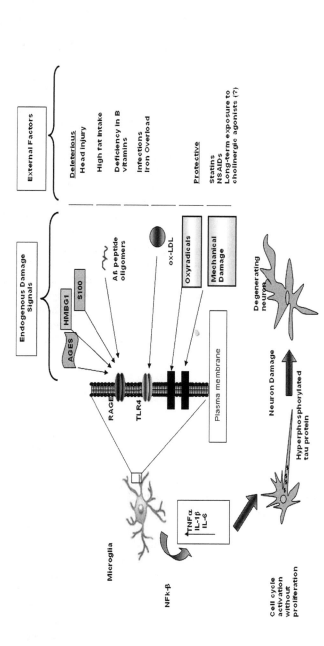

Fig. 1 Schematic representation of the hypothetical roles of endogenous danger/alarm signals built into the innate immune system at the early stages of the pathogenesis of Alzheimer's disease (*AD*). Danger signals such as advanced glycation end products (*AGES*), high mobility box group 1 (*HMBG1*), S-100 proteins, and Aβ peptide oligomers (but not β-pleated fibrillar aggregates) activate microglia through the AGES receptor (RAGE; shown here as a transmembrane protein in a cartoon of a lipid bilayer). Separately, oxidized low-density lipoproteins (*oxLDL*) activate toll-like receptors (*TLRs*), particularly TLR4. Additional danger signals, such as trauma and oxyradical damage, possibly acting on separate receptors (*black boxes* inserted in the membrane) as well as by inducing the production of additional Aβ peptide oligomers, AGES, and S-100 protein contribute to this process. Separately and in various combinations (as it may apply to different individuals), these danger signals would trigger innate immune system alarm mechanisms resulting in long-term excessive levels of tumor necrosis factor-α (*TNFα*), interleukin 1-β (*IL1-β*), and IL-6. These signals would then mediate neuronal damage directly, reflected in alterations such as tau protein hyperphosphorylation and paired helical filaments formation, which eventually result in neuronal degeneration, interstitial Aβ peptide aggregation, and its deposition in a β-pleated configuration, and progressively more severe clinical manifestations of cognitive and behavioral decline unrelated to the amyloid aggregation process (Reproduced with permission from Fernandez et al., 2008) (*See Color Plates*)

In the second part of this chapter, we critically analyze the influences of oxidative stress, cholesterol, and lipids in the molecular mechanisms of AD pathogenesis, as well as their potential relevance for early AD diagnosis. An increasing number of reports link altered cholesterol metabolism and other lipid abnormalities with the pathogenesis of AD [4–6]. Interestingly, "statins," which are inhibitors of the 3-hydroxy-3-methylglutaryl-coenzyme A reductase (HMGCo-A reductase), the enzyme that catalyzes the early rate-limiting reaction in cholesterol biosynthesis, seem to protect against dementia and AD [7, 8], suggesting that a reduction in the levels of systemic cholesterol could delay the development of cognitive decline.

However, "statins" are not only inhibitors of the HMGCo-A reductase but also have pleiotropic effects, which are independent from changes in cholesterol biosynthesis. Therefore, the overall cognitive benefits observed by using "statins" appear to be greater than what might be expected only from changes in the cholesterol levels. Indeed, recent studies indicate that some of the pleiotropic effects of statins also involve an improvement in endothelial function, enhancing the stability of atherosclerotic plaques, decreasing oxidative stress and inflammation, and inhibiting the thrombogenic response [7]. Many of these pleiotropic effects are mediated by the inhibition of the isoprenoids pathway. Isoprenoids serve as lipid attachments for intracellular signaling molecules such as the GTP-binding proteins, Rho, Ras, and Rac, whose proper membrane localization and function depend on isoprenylation. Also, the peroxisome proliferator-activated receptor-α activation by "statins" appears to be relevant for the overall effect on cognition [7]. Therefore, alterations in cholesterol metabolism may shed light on the search for biomarkers for AD.

2 Inflammatory Cytokines and AD Onset

Glial cells normally provide neurotrophic factors essential for neurogenesis, and when activated by a set of stressing events, they overproduce cytokines and NGF, thus triggering altered signaling patterns that appear to contribute to the pathogenesis of AD [1]. A set of discoveries has strengthened the idea that altered patterns in the glia–neuron interactions constitute early molecular events within the cascade of cellular signals that lead to neurodegeneration [5, 9–11]. Indeed, brain inflammation is responsible for the abnormal secretion of proinflammatory cytokines that trigger neuronal tau hyperphosphorylation in residues that are not modified under normal physiological conditions [12]. However, other cytokines such as IL-3 and TGFβ1, which seem to display neuroprotective activities rather than deleterious effects [13], are decreased in mononuclear cells in cognitively impaired patients. Thus, the family of cytokines seems to display dual effects on neuronal degeneration: under specific conditions, some of them, for example, IL-6, IL-1, and TNFα trigger the neurodegenerative cascade, while others (IL-3 and TGFβ1) protect against neuronal damage.

In order to understand the implications of proinflammatory cytokines for the molecular diagnosis of AD, it is critical to address two main questions: (1) What are the relationships between the mechanisms of cytokine control and inflammation of the CNS? and (2) What are the relationships between peripheral cytokines levels and the cognitive decline in humans? The first question has been analyzed above, while the second one is discussed below.

3 Cytokines Levels in Peripheral Tissues and CNS

The mechanisms by which peripheral immune mediator levels affect brain functioning, regulation of cytokines in the CNS, and peripheral tissues have not been completely understood yet. During the early 1990s, this field of neuroimmunology was dominated by the paradigm that the peripheral and central immune system cannot signal each other, since the large molecules such as cytokines (15- to 20-kDa proteins) do not cross the blood–brain barrier (BBB) [14]. Today, the evidence around this topic indicates that the amount of cytokines that can enter the brain is modest but comparable to other small soluble compounds (<300 Da), such as morphine, known to cross the BBB in sufficient amounts to affect brain function [15]. More importantly for AD, efflux of cytokines from CNS to blood has also been shown to occur, specifically for IL-1β, IL-6, and TNFα [16, 17], but the mechanisms of passage for these cytokines and its implications for AD diagnosis are still unclear. However, many reports describe pathways for peripheral cytokines to enter the CNS, which are important in order to understand potential mechanisms of brain influx of cytokines. There are at least three known pathways by which peripheral cytokines exert their effects on the brain: (1) peripheral tissues, innervated by the peripheral and autonomic nervous systems, direct signals to the brain via peripheral nerves; (2) brain vasculature convey signals through secondary messengers, such as nitric oxide (NO) or prostanoids, produced in response to cytokines; and (3) cytokines that can directly act at the level of the brain parenchyma after crossing BBB or after entering brain areas that lack a BBB [17]. Interestingly, BBB permeability for cytokines can be increased by NO delivered from local microglial cells, which is relevant considering that inducible NOS (iNOS) is increased in leukocytes from AD patients [3]. Cytokines can also be carried across the BBB by active transport. Such transport is saturable and specific for individual cytokines, and does not apply to all cytokines. Thus, it seems unlikely that acute changes in CNS cytokine levels result in dramatic modifications of cytokines in peripheral fluids unless the mechanisms of BBB permeability and specific transporters for cytokines become damaged. These saturable mechanisms are not modified by anti-inflammatory agents such as dexamethasone or nonsteroidal anti-inflammatory drugs (NSAIDs) such as indomethacin. Therefore, the administration of NSAIDs in AD patients seems not to affect cytokine efflux from brain. Thus, the levels of cytokines in any peripheral biological fluid result not only from the brain-peripheral blood efflux but as a result of the net exchange of cytokines from two anatomically and physiologically different compartments separated by the BBB.

Significant increases of TNFα and IL-6 in plasma and cerebrospinal fluid (CSF) and decreased levels of plasma TGFβ1 and IL-3 have been found in the early stages of AD, suggesting that those can be potential biological markers. The sCD40 was shown to be significantly increased in the plasma of AD patients [18], while TGFβ1 is significantly decreased in plasma of those patients [19]. Leukocytes NOS is also a potential marker that is significantly increased in AD patients.

A recent epidemiological study in elderly population has shown that individuals with higher peripheral IL-6 levels exhibit higher rates of cognitive decline over a period of 7 years [20]. Also, a promising role has been attributed to TNFα and IL-1β in the detection of AD [9, 21]. Indeed, the perispinal injection of etanercept, a TNFα inhibitor widely used for the treatment of rheumatoid arthritis, provided some improvements in cognitive function for 15 AD patients [22]. The relevance of this particular study is still being evaluated because of the small population of subjects in the trials and the facts that no controls are shown. Taken together, all these studies point to a major role of TNFα (*see* Fig. 1, not only in the molecular signals that lead to neurodegeneration [1, 5, 10] but also in the clinical course of AD.

The Longitudinal Aging Study Amsterdam [23], one of the most consistent and well-designed clinical studies performed to address the potential role of serum C-reactive protein, IL-6, and albumin in the diagnosis of AD (1,284 patients, aged 62–85 years), did not detect any association between poor memory performance or slow information processing and cognitive impairment, with the serum levels of IL-6; only the serum inflammatory protein α1-antichymotrypsin was associated with cognitive decline in older persons [23]. However, the interpretation of these findings is seriously limited by the fact that brain accumulation of cytokines has been shown to occur during acute phases of the disease, and as mentioned above, not all proteins efficiently pass through the BBB. Again, it is worth notice that since serum is in a completely different biological compartment than CNS, changes in the levels of soluble molecules such as TNFα, IL-6, IL-3, TGFβ1, and sCD40 can be limited by saturable mechanisms of transport, thus acute changes in concentrations of cytokines in brain might not be detected in serum. Another important issue to be considered in the analysis of cytokines in biological fluids is the limit of detection (LOD), which is the lowest concentration of a certain biomarker that can be detected with acceptable precision and accuracy (Table 1). ELISA systems are the most popular methods for the determination of cytokines in biological samples; however, sensitivities of methods vary and LODs seem to be, in many cases, above the concentration of the cytokines in normal and AD patients [16]. This can be a serious problem when concentrations of the cytokines, for example, IL-6, are around 0.5 pmol/liter [16]. Thus, the discrepancies from different studies when comparing AD patients and controls are explained in part by different sensitivities in the methods of biochemical analysis. More important for correlation studies between cytokine concentrations and cognitive decline is the limit of quantification (LOQ), which is the limit at which any laboratory can reasonably tell the difference between two different concentration values. As LOQ can be drastically different between laboratories, sometimes LOQ is defined simply as about five times the LOD.

Unfortunately, in spite of the great amount of evidence collected up to date on the involvement of proinflammatory cytokines in human cognitive decline, none of these immunological biomarkers has been introduced as a routine diagnostic tool

Table 1 Controversial results in the assessment of proinflammatory cytokines in serum and CSF of demented patients and controls

| Sample | Marker | |
	Levels in respect to control	References
CFS	IL-6	
	+	[40–42]
	=	[43, 44]
	IL1β	
	+	[45]
	=	[43, 44]
	TNFα	
	+	[43]
	=	[44]
Serum	IL-6	
	+	[20]
	−	[46]
	=	[23]
	IL1β	
	+	[18]
	=	[45]
	TNFα	
	+	[47, 48]
	−	[49]
	sCD40	
	+	[18]
	TGFβ	
	−	[19, 50]
	IL-3	
	−	[50]
	iNOS	
	+	[3]

CSF cerebrospinal fluid, *IL-6* interleukin-6, *TNFα* tumor necrosis factor-α, *sCD40* soluble CD40, *TGFβ* tumoral growth factor-β, *iNOS* inducible nitric acid synthase
+ Increased levels in AD compared to control
− Decreased levels in AD compared to control
= No differences in levels of AD patients compared to control

for AD, or even as a predictor for eventual dementia. An important reason for this could be the discrepancies between clinical studies, based on the practical issues, for example, different sensitivity of the analytical methods or major design problems. For example, it is difficult to design clinical studies to assess concentrations of cytokines in CSF of healthy patients due to ethical restrictions. Some studies that compare cytokine levels in CSF samples have been developed in AD patients, but using controls such as mild cognitive impairment (MCI), Parkinson, and other non-AD demented patients [24]. These limitations can delay the search for a reliable biomarker for early stages of AD.

In the light of the controversy on cytokine levels in biological samples (Table 1), it seems reasonable to consider that polymorphic forms of the proinflammatory cytokines might have a role in the clinical course of AD. So far, very little information exists on the prevalence of polymorphisms of inflammatory cytokines among normal and demented subjects. The relationship between specific IL-6 genotypes and cognitive function was recently shown in a study on newborns, and the IL-6-572 C allele (CC/GC genotypes) was associated with cognitive impairment linked to embryogenesis and brain development [25]. Interestingly, the cognitive function in IL-6 knockout (KO) mice is better than in wild-type animals. The IL-6 KO animals showed better performance in various cognitive tests, as compared with wild-type mice. Again, these basic studies strongly support the idea that IL-6, a proinflammatory cytokine, plays an important role in memory processing [26]. In addition, it has been shown that individuals homozygous for the IL-1α −889*1 allele decline more rapidly on the Mini-Mental State Examination than other demented patients [27], suggesting that this polymorphism renders individuals more vulnerable to dementia. In another set of studies, IL-1β (1418C>T) and TNFα (308G>A) gene polymorphisms induce protein expression of the corresponding serum cytokines affecting human cognitive performance [9]. These findings are supported by studies showing dramatic alterations on memory processing by proinflammatory cytokines [28]. Moreover, these findings support the hypothesis that inflammation is important in the clinical course of AD and more specifically, that polymorphisms of cytokines can dramatically affect the cognitive decline of demented patients. Again, the overall scientific data suggest that an equation that combines different markers, for example, polymorphisms, cytokines levels associated to CSF concentrations of amyloid-β (Aβ) peptide, or phosphorylated tau, appears as the most appropriate tool to monitor the clinical course of demented patients.

4   Cholesterol and Lipids in AD

The search for a peripheral biomarker based on cholesterol metabolism must consider the structural features of the main cholesterol and lipids variants in the bloodstream. Thus, lipids and cholesterol are transported in the bloodstream by lipoprotein particles that can be classified in different subsets according to their density: high density (HDL), medium density (IDL), low density (LDL), and very low density (VLDL). These proteins have been widely used as a tool to assess the cardiovascular risk. Up to date, different roles have been found for each of these lipoproteins. LDL, which has an elevated fat proportion, participates in lipids transported from blood stream to peripheral tissues. HDL, with a major protein fraction, transports cholesterol from peripheral tissues to the liver where conjugation and excretion occur. Interestingly, in the last 5 years, increasing numbers of reports have described the putative relationships among cholesterol, lipoproteins, and cognitive impairment [29–31].

Hypercholesterolemia is an early risk factor for the development of amyloid pathology and longitudinal, population-based studies have demonstrated that abnormal cholesterol levels are associated with AD in the later life span [32]. However, the mechanisms by which cholesterol and other lipids affect cognitive performance are still an open field of research. One of the challenges of correlating the systemic lipid biomarkers with brain cholesterol is that there is virtually no transfer of peripheral cholesterol to the brain. Brain cholesterol is synthesized mainly in the CNS. It is estimated that during CNS development, neurons synthesize most of the cholesterol needed for their growth and synaptogenesis. Later, when neurons mature, they reduce their endogenous cholesterol synthesis and become more dependent on cholesterol synthesized and secreted by astrocytes [5]. Cholesterol in the CNS is turned over in a proportion of 0.7% of the total amount every day. Even though it is not a substantial fraction, it is a relatively high amount of cholesterol considering that the CNS accounts for 2.1% of body weight and contains 23% of the total sterols in the body. The brain is therefore the most cholesterol-rich human organ. In addition, cholesterol accounts for 20–25% of the total lipids in neuron plasma membranes. Thus, neurons require a continuous supply of new cholesterol to maintain a constant concentration in plasma membranes. The mechanisms by which eukaryotic cells incorporate cholesterol are at least three: (1) de novo synthesis within the cell from acetyl-CoA, which is the most important mechanism for neurons; (2) uptake of unesterified or esterified cholesterol from the external environment using the LDL receptors; or (3) the Niemann-Pick C1-like protein (NPC1L1).

Alterations of cholesterol metabolism, for example, hypercholesterolemia, and redox imbalance linked to normal and pathological cognitive decline [33]. This has encouraged the search of biomarkers for the initial phases of AD based on changes in lipid profile and redox biomarkers (34, 35, 36). Among them are: 4-hydroxyno-neal (4-HNE), a lipoperoxide produced by the oxidation of arachidonic acid, and 24S-hydroxycholesterol, a metabolite produced via enzymatic oxidation of brain cholesterol by CYP46, in plasma and urine (Fig. 2). The assessment of these lipid-derived biomarkers indicates that 4-HNE in plasma and CSF levels of 24S-hydroxycholesterol are elevated in AD patients [34]. Also malondialdehyde, an end product of lipid peroxidation, has been reported to be increased in the plasma of AD patients versus age-matched and nutritionally evaluated control subjects [35]. Nevertheless, malondialdehyde failed as a predictive AD marker in several other studies [31, 35].

Interestingly, levels of 24S-hydroxycholesterol have shown to be increased in plasma of AD patients compared with control age-matched volunteers [37] (Fig. 2). Considering that the treatment with statins lowers the levels of LDL-cholesterol and 24S-hydroxycholesterol in plasma [4], the assessment of oxysterol has become an important biomarker for AD risk more than other forms of cholesterol.

Other lipids that have also been assessed as potential biomarkers for AD are the sulfatides, which are a class of sulfated galactocerebrosides that mediate diverse biological processes including cell growth regulation, protein trafficking, signal transduction, adhesion, neuronal plasticity, and cell morphogenesis. These studies

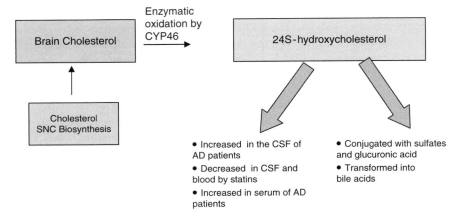

Fig. 2 Brain cholesterol is mostly synthesized in the central nervous system. 24S-Hydroxycholesterol is a metabolite produced via enzymatic oxidation of brain cholesterol and is eliminated as sulfatides and glucuronides or transformed into bile acids. Blood levels of this cholesterol metabolite are normally decreased in elderly and abnormally increased in the serum and cerebrospinal fluid (*CSF*) of Alzheimer's disease (*AD*) patients (*See Color Plates*)

show that sulfatides' concentration decrease in very mild dementia patients [38], but a correlation with AD has not been established yet.

Our group has searched for associations of LDL, HDL, VLDL, and total levels of cholesterol in plasma from AD patients with cognitive impairment and healthy age-matched controls and found apparently no significant correlations for the majority of the lipoproteins we studied [31]. This is consistent with recent reports describing no association between a variety of lipids and the risk of amnestic or nonamnestic MCI [37]. However, we have found a slight negative correlation between Geriatric Depression Scale (GDS) scores against total cholesterol, also GDS status against VLDL levels. More importantly, we have reported that there is a positive correlation between the cholesterol/HDL ratio and the impairment of semantic memory by Boston Naming Test. The latter is a very interesting finding in the light of recent reports describing correlations between metabolic syndrome and a high risk of cognitive decline [39]. Thus, the assessment of cholesterol/HDL ratio in patients with metabolic syndrome may offer a potential marker for cognitive impairment in this specific group of patients.

The relationships between cholesterol metabolism and normal and pathological cognitive decline are still unclear; however, it seems that combinations between redox and lipid biomarkers offer a very tempting avenue for future diagnostic tools of AD and other type of dementias, for example, 24S-hydroxycholesterol, 4-HNE plasma levels, to date good candidates to be markers of AD. MDA (malondialdehyde) has been assessed in several studies in different groups of normal and cognitively impaired patients but no correlation has been found yet for this marker.

The alterations in levels of plasma lipoproteins seem to have no correlation with different types of cognitive impairment, yet combined markers such as cholesterol/HDL ratio [31] seem to offer interesting new avenues in the future search for AD biomarkers based on alterations of the systemic lipid profile.

Acknowledgments Research was supported by the International Center for Biomedicine (ICC), FONDECYT grant 1080264, and by a grant from Arturo Prat University to LR.

References

1. Fernandez JA, Rojo L, Kuljis RO, Maccioni RB (2008) The damage signals hypothesis of Alzheimer's disease pathogenesis. J Alzheimers Dis 14(3):329–333
2. Rojo LE, Fernández JA, Maccioni AA, Jimenez JM, Maccioni RB (2008) Neuroinflammation: implications for the pathogenesis and molecular diagnosis of Alzheimer's disease. Arch Med Res 39:1–16
3. Deservi B, La Porta CA, Bontempelli M, Comolli R (2002) Decrease of TGF-b1 plasma levels and increase of nitric oxide synthase activity in leukocytes as potential biomarkers of Alzheimer's disease. Exp Gerontol 37:813–821
4. Locatelli S, Lutjohann D, Schmidt HH-J, Otto C, Beisiegel U, von Bergmann K (2002) Reduction of plasma 24S-hydroxycholesterol (cerebrosterol) levels using high-dosage simvastatin in patients with hypercholesterolemia: evidence that simvastatin affects cholesterol metabolism in the human brain. Arch Neurol 59:213–216
5. Rojo L, Sjöberg MK, Hernández P, Zambrano C, Maccioni RB (2006) Roles of cholesterol and lipids in the etiopathogenesis of Alzheimer's disease. J Biomed Biotechnol 2006:73976
6. Reid PC, Urano Y, Kodama T, Hamakubo T (2007) Alzheimer's disease: cholesterol, membrane rafts, isoprenoids and statins. J Cell Mol Med 11:383–392
7. Lio JK, Laufs U (2005) Pleiotropic effects of statins. Annu Rev Pharmacol Toxicol 45:89–118
8. Wolozin B, Kellman W, Ruosseau P, Celesia GG, Siegel G (2000) Decreased prevalence of Alzheimer disease associated with 3-hydroxy-3-methylglutaryl coenzyme A reductase inhibitors. Arch Neurol 57:1439–1443
9. Lio D, Annoni G, Licastro F, et al. (2006) Tumor necrosis factorα-308A/G polymorphism is associated with age at onset of Alzheimer's disease. Mech Ageing Dev 127:567–571
10. Quintanilla RA, Orellana DI, González-Billault C, Maccioni RB (2004) Interleukin-6 induces Alzheimer-type phosphorylation of tau protein by deregulating the cdk5/p35 pathway. Exp Cell Res 295:245–257
11. Orellana DI, Quintanilla RA, Gonzalez-Billault C, Maccioni RB (2005) Role of the JAKs/STATs pathway in the intracellular calcium changes induced by interleukin-6 in hippocampal neurons. Neurotox Res 8:295–304
12. McGeer PL, Rogers J, McGeer EG (2006) Inflammation, anti-inflammatory agents and Alzheimer disease: the last 12 years. J Alzheimers Dis 9 (3 Suppl):271–276
13. Zambrano CA, Egaña JT, Núñez MT, Maccioni RB, González-Billault C (2004) Oxidative stress promotes tau dephosphorylation in neuronal cells: the roles of cdk5 and PP1. Free Radic Biol Med 36:1393–1402
14. Bocci V (1998) Central nervous system toxicity of interferons and other cytokines. J Biol Regul Homeost Agents 2:107–118
15. Licinio J, Wong ML (1997) Pathways and mechanisms for cytokine signaling of the central nervous system. J Clin Invest 100:2941–2947
16. Teunissen CE, de Vente J, Steinbusch HW, De Bruijn C (2002) Biochemical markers related to Alzheimer's dementia in serum and cerebrospinal fluid. Neurobiol Aging 23:485–508

17. Banks WA, Kastin AJ (1997) Relative contributions of peripheral and central sources to levels of IL-1a in the cerebral cortex of mice: assessment with species-specific enzyme immunoassays. J Neuroimmunol 79:22–28
18. Galasko D (2005) Biomarkers for Alzheimer's disease clinical needs and application. J Alzheimers Dis 8:339–346
19. Mocali A, Cedrola S, Della Malva N, et al. (2004) Increased plasma levels of soluble CD40, together with the decrease of TGF beta 1, as possible differential markers of Alzheimer disease. Exp Gerontol 39:1555–1561
20. Weaver JD, Huang MH, Albert M, Harris T, Rowe JW, Seeman TE (2002) Interleukin-6 and risk of cognitive decline: MacArthur studies of successful aging. Neurology 59:371–378
21. Sciacca FL, Ferri C, Licastro F, et al. (2003) Interleukin-1B polymorphism is associated with age at onset of Alzheimer's disease. Neurobiol Aging 24:927–931
22. Griffin W (2008) Perispinal etanercept: potential as an Alzheimer therapeutic. J Neuroinflammation 5:3–5
23. Comijs HC, Dik MG, Aartsen MJ, Deeg DJ, Jonker C (2005) The impact of change in cognitive functioning and cognitive decline on disability, well-being, and the use of healthcare services in older persons. Results of Longitudinal Aging Study Amsterdam. Dement Geriatr Cogn Disord 19:5–6
24. Maccioni RB, Lavados M, Maccioni CB, Farias G, Fuentes P (2006) Anomalously phosphorylated tau protein and Abeta fragments in the CSF of Alzheimer's and MCI subjects. Neurobiol Aging 27:237–244
25. Harding D, Brull D, Humphries SE, Whitelaw A, Montgomery H, Marlow N (2005) Variation in the interleukin-6 gene is associated with impaired cognitive development in children born prematurely: a preliminary study. Pediatr Res 58:117–120
26. Braida D, Sacerdote P, Panerai AE, et al. (2004) Cognitive function in young and adult IL (interleukin)-6 deficient mice. Behav Brain Res 153:423–429
27. Murphy GM Jr, Claassen JD, DeVoss JJ, et al. (2001) Rate of cognitive decline in AD is accelerated by the interleukin-1 alpha–889 *1 allele. Neurology 56:1595–1599
28. Pollmacher T, Haack M, Schuld A, Reichenberg A, Yirmiya R (2002) Low levels of circulating inflammatory cytokines, do they affect human brain functions? Brain Behav Immun 16:525–532
29. Dimopoulos N, Piperi C, Salonicioti A, et al. (2007) Characterization of the lipid profile in dementia and depression in the elderly. J Geriatr Psychiatry Neurol 20:138–144
30. Yaffe K (2007) Metabolic syndrome and cognitive decline. Curr Alzheimer Res 4:123–126
31. Sekler A JJ, Leonel E Rojo, Edgar pastene, Patricio Fuentes, Andrea Slachevsky, Ricardo B Maccioni (2008) Cognitive Impairment Associated with Alzheimer's Disease: Links with Cholesterol Metabolism and Oxidative Stress. Neuropsychiatric Disease and Treatment 4: 4(4):715–722
32. Kivipelto M, Helkala EL, Laakso MP, et al. (2001) Midlife vascular risk factors and Alzheimer's disease in later life: longitudinal, population based study. BMJ 322:1447–1451
33. Sparks DL, Sabbagh MN, Breitner JC, Hunsaker JC 3rd; AD Cholestrol-Lowering Treatment Trial Team and the Ancillary ADAPT: Cholestrol and Statin Parameters Work Group; Cache County and ADAPT Work Groups; Eastern Division of the Kentucky Medical Examiner's Group (2003) Is cholesterol a culprit in Alzheimer's disease? Int Psychogeriatr 15 Suppl 1:153–159
34. Selley ML, Close DR, Stern SE (2002) The effect of increased concentrations of homocysteine on the concentration of (E)-4-hydroxy-2-nonenal in the plasma and cerebrospinal fluid of patients with Alzheimer's disease. Neurobiol Aging 23:383–388
35. Bourdel-Marchasson I, Delmas-Beauvieux M-C, Peuchant E, et al. (2001) Antioxidant defences and oxidative stress markers in erythrocytes and plasma from normally nourished elderly Alzheimer patients. Age Ageing 30:235–241
36. Reitz C, Tang MX, Manly J, Schupf N, Mayeux R, Luchsinger JA (2008) Plasma lipid levels in the elderly are not associated with the risk of mild cognitive impairment. Dement Geriatr Cogn Disord 25:232–237

37. Lutjohann D, Papassotiropoulos A, Bjorkhem I, et al. (2000) Plasma 24S-hydroxycholesterol (cerebrosterol) is increased in Alzheimer and vascular demented patients. J Lipid Res 41:195–198
38. Han X, Fagan AM, Cheng H, Morris JC, Xiong C, Holtzman DM (2003) Cerebrospinal fluid sulfatide is decreased in subjects with incipient dementia. Ann Neurol 54:115–119
39. Komulainen P, Lakka TA, Kivipelto M, et al. (2006) Metabolic syndrome and cognitive function: a population-based follow-up study in elderly women. Dement Geriatr Cogn Disord 23:29–34
40. Blum-Degen D, Muller T, Kuhn W, Gerlach M, Przuntek H, Riederer P (1995) IL-1 beta, and IL-6 are elevated in the cerebrospinal fluid of Alzheimer's, and de novo Parkinson's disease patients. Neurosci Lett 202:17–20
41. Martinez M, Fernandez-Vivancos E, Frank A, De la Fuente M, Hernanz A (2000) Increased cerebrospinal fluid fas (Apo-1) levels in Alzheimer's disease. Relationship with IL-6 concentrations. Brain Res 869:216–219
42. Martinez M, Frank A, Hernanz A (1993) Relationship of interleukin-1 beta and beta 2-microglobulin with neuropeptides in cerebrospinal fluid of patients with dementia of the Alzheimer type. J Neuroimmunol 48:235–240
43. Tarkowski E, Blennow K, Wallin A, Tarkowski A (1999) Intracerebral production of TNF-alpha, a local neuroprotective agent in Alzheimer disease and vascular dementia. J Clin Immunol 19:223–230
44. Lanzrein AS, Johnston CM, Perry VH, Jobst KA, King EM, Smith AD (1998) Longitudinal study of inflammatory factors in serum, cerebrospinal fluid, and brain tissue in Alzheimer disease: IL-1beta, IL-6, IL-1 receptor antagonist, TNF-alpha, the soluble TNF receptors I and II, and alpha1 – antichymotrypsin. Alzheimer Dis Assoc Disord 12:215–227
45. Cacabelos R, Franco-Maside A, Alvarez XA (1991) Interleukin-1 in Alzheimer's disease and multi-infarct dementia: neuropsychological correlations. Methods Find Exp Clin Pharmacol 13:703–708
46. Singh VK, Guthikonda P (1997) Circulating cytokines in Alzheimer's disease. J Psychiatr Res 31:657–660
47. Jia JP, Meng R, Sun YX, Sun WJ, Ji XM, Jia LF (2005) Cerebrospinal fluid tau, Abeta1–42 and inflammatory cytokines in patients with Alzheimer's disease and vascular dementia. Neurosci Lett 383:12–16
48. Alvarez A, Cacabelos R, Sanpedro C, García-Fantini M, Aleixandre M (2007) Serum TNF-alpha levels are increased and correlate negatively with free IGF-I in Alzheimer disease. Neurobiol Aging 28:533–536
49. Cacabelos R, Alvarez XA, Franco-Maside A, Fernandez-Novoa L, Caamano J (1994) Serum TNF in Alzheimer's disease, and multi-infarct dementia. Methods Find Exp Clin Pharmacol 16:29–35
50. Shalit F, Sredni B, Stern L, Kott E, Huberman M (1994) Elevated interleukin-6 secretion levels by mononuclear cells of Alzheimer's patients. Neurosci Lett 174:130–132

Participation of Glial Cells in the Pathogenesis of AD: A Different View on Neuroinflammation

Rommy von Bernhardi

Abstract Inflammation has been linked to Alzheimer's disease (AD). The accepted view is that inflammation is secondary to amyloid-β (Aβ) accumulation or neuro-degeneration. However, I propose that glial dysfunction and the resulting imbalance between the cytotoxic inflammatory and neuroprotective-modulator activity could be the pathological mechanism behind AD. Such unbalance could be promoted by conditions like hypoxia and inflammation, which are frequently observed in aged individuals. A strong inflammatory response can promote defective processing of the amyloid-β protein precursor (AβPP) and the handling of Aβ by glial cells, resulting in the accumulation of Aβ and further inflammation. Proinflammatory conditions also enhance microglial cell activation by AβPP and Aβ and reduce astrocytes-mediated inhibition of microglial activation. These observations indicate that glial cell response to Aβ can be critically dependent on the priming of glial cells by proinflammatory factors. Astrocytes play a major role in the pathophysiology of AD, both promoting damage and mediating neuroprotection. Persistent inflammation can impair modulation and promote microglia-mediated neurotoxicity. Altogether, I propose that dysfunctional glia could result in both neuroinflammation and impaired neuronal function in AD.

Abbreviations Aβ, amyloid-β; AβPP, amyloid-β protein precursor; AD, Alzheimer's disease; BACE 1, β-secretase cleaving enzyme 1; CNS, central nervous system; COX, cyclooxygenase; COX-2, inducible form of cyclooxygenase; ERK, extracellular signal-regulated kinases; IFN-γ, interferon-γ IL-1β, interleukin-1β; IL-6, interleukin-6; iNOS, inducible nitric oxide synthase; JNK, jun N-terminal kinase; LPS, lipopolysaccharide; LTP, long-term potentiation; MAPKs, mitogen-activated protein kinases; MCP-1, monocyte chemotactic protein-1; MHC-class

R. von Bernhardi (✉)
Department of Neurology
Faculty of Medicine, Pontificia Universidad Católica de Chile
Marcoleta 391, Santiago, Chile
e-mail: rvonb@med.puc.cl

R.B. Maccioni and G. Perry (eds.) *Current Hypotheses and Research Milestones in Alzheimer's Disease*. DOI: 10.1007/978-0-387-87995-6_12,
© Springer Science + Business Media, LLC 2009

II, major histocompatibility complex class II; MKP-1, MAPK phosphatase type 1; NF-κB, nuclear factor-κ B; NGF, nerve growth factor; NO, nitric oxide; NSAIDs, nonsteroidal anti-inflammatory drugs; ROS, reactive oxygen species; SRs, scavenger receptors; STAT1, signal-transducer and activator of transcription-1; TGF-β, transforming growth factor-β; TNF-α, tumor necrosis factor-α

1 Introduction

Alzheimer's disease (AD) is a neurodegenerative disorder characterized by impaired cognitive functions associated with neuronal dysfunction and selective neuronal loss. Its neuropathology is characterized by two lesions: the extracellular aggregation of fibrillar amyloid-β (Aβ) in senile plaques and intracellular neurofibrillary tangles. Both lesions are closely associated with abundant activated microglia and astrocytes. However, whether amyloid plaques and neurofibrillary tangles are the primary cause of AD or a consequence of another pathophysiological change (i.e., the proinflammatory environment associated with aging) is still an open question. There is a lack of correlation between the extent of amyloid pathology and clinical progression. The accumulation of Aβ does not necessarily constitute a senile plaque. Diffuse cortical plaques are frequently observed in cognitively preserved elders, but they do not show glial and inflammatory reactivity. In contrast, senile plaques in AD are fibrillar and associate with activated glia and an inflammatory response [1]. Furthermore, there is no clear correlation between amyloid deposition and the regions with most of the neuronal loss in early AD. Such differences make it difficult to argue that just the exposure to Aβ directly causes neurodegeneration and suggest that the deposition of Aβ in the brain is not necessarily associated to functional impairment or cell death. Additional pathogenic events are needed for Aβ to accumulate as well as to become cytotoxic, and they could be provided by glial cells. Here I will discuss evidence supporting the role of glial dysfunction in the pathogenesis of AD, including aging-associated changes, modulation of glial cell function, inflammatory cytokines, and cell receptors that could perpetuate neuroinflammation through Aβ-mediated cell activation.

2 Age-Dependent Changes Relevant for AD

The strongest risk factor for AD is aging. Several changes relevant to AD occur in aging. Autopsy studies show that the incidence of neuropathological lesions is even higher than clinical symptoms, and reveal that AD often occurs in conjunction with other pathological lesions, especially vascular and Lewy body dementia. The overlap of pathologies suggests the existence of common pathophysiological mechanisms. Age-related changes include increased reactivity of microglial cells, cytokine secretion, increased production of nitric oxide (NO), and oxidative stress secondary to increased production of reactive oxygen species (ROS). It is likely that several of the resulting changes, oxidative changes of proteins and lipids, increased glycation, calcium and energetic metabolism among others induce glial cell activation and upregulation of inflammatory mediators.

Brain hypoperfusion [2] affects brain parenchyma homeostasis resulting in disruption of energetic metabolism and the maintenance of ionic gradients, increased intracellular calcium, and production of oxygen free radicals. Cerebral ischemia induces a strong cerebral inflammatory response, upregulation of amyloid-β protein precursor (AβPP) expression, and increased amyloidogenic cleavage of AβPP. Furthermore, in addition to age-dependent hypoperfusion, the microcirculation in AD patient brains and AβPP transgenic mice have abnormalities which can further promote ischemia and hypoxia. Furthermore, transient hypoxia can lead to mitochondrial dysfunction, also resulting in oxidative stress, impairment of membrane integrity, and amyloidogenic cleavage of AβPP with a pattern similar to that observed in sporadic AD.

Expression of several receptors changes with aging. Upregulation of CD14 could be associated with increased inflammatory reactivity in normal aging [3]. Similar changes are also observed in AβPP transgenic mice [4]. Microglial cells from aged individuals secrete several of the cytokines upregulated in microglial cells exposed to Aβ. These observations plus the morphological abnormalities described in microglia of cognitively normal elders led to the concept of dystrophic microglia associated with aging [5].

Impairment of the removal of Aβ by microglia could also be a consequence of age-related changes. Glial cells express various receptors that bind Aβ under normal and pathological conditions. We have reported that microglia effectively bind to and phagocytes Aβ in vitro [6], whereas they do not appear to eliminate Aβ aggregates in vivo [7] or under proinflammatory conditions. In contrast, newly differentiated microglial cells originating from the blood are able to remove Aβ from the brain parenchyma. Animals that fail to recruit microglia have more plaques [8]. These results suggest that microglia are less immune competent than their blood-derived counterparts.

Aging, as well as AD, associates to change in expression pattern of scavenger receptors (SRs). Several of them appear to be receptors for Aβ [6, 9]. SRs can be especially relevant because of their participation in the uptake of Aβ as well as mediators of glial inflammatory activation [9]. Upregulation of SRs in blood-derived macrophages associates with promotion of cell survival [10]. The binding of Aβ induces the expression of some of the SRs, the secretion of chemokines and proinflammatory cytokines, increased production of ROS, and the activation of various signaling transduction pathways. Microglia express many SRs of various types in agreement with their function as scavenger cells. In contrast, the more restricted expression of SRs by astrocytes could result in a different activation response. Impairment of Aβ clearance and the activation of signal transduction pathways by SRs binding Aβ could be involved in the inflammatory response and participate in Aβ accumulation.

On a different front, synaptic function is attenuated and the ability to generate LTP is nearly abolished in aged rats. Aged monkeys exhibit cortical neuritic pathology, amyloid plaques, and reactive glia associated with senile plaques. The response to Aβ also changes with aging. Aβ induces neuronal loss and microglial reactivity only when injected in aged monkeys. Considering that aging changes degradation of Aβ by astrocytes [11] and is associated with increased oxidative potential of Aβ, one can speculate that Aβ accumulation and its potential toxic effect through glial inflammatory activation are potentiated in the aged brain.

3   Damage Mechanisms: Aβ Versus Inflammation

The quest for finding a unique entity responsible for sporadic AD probably will fail because multiple biological mechanisms appear to be involved. A wealth of evidence suggests that inflammatory processes can be responsible for neuronal dysfunction and degeneration. Resident cells such as microglia and astrocytes secrete cytokines, chemokines, proteases, complement proteins, and ROS. However, there is also compelling evidence indicating that inflammatory cells and mediators also have beneficial functions in the central nervous system (CNS), being also necessary for the accomplishment of several normal processes. The dual role for glia-mediated neuroinflammation suggests that glial cells dysregulation through the impairment of their fine-tuned multiple functions may be involved in the genesis of AD [12].

Several mechanisms have been implicated in Aβ cytotoxicity for neurons and glial cells. Aβ has direct neurotoxicity but it also kills neurons indirectly through the activation of microglial cells [13]. Indirect cytotoxicity appears to be as important as the direct mechanism (Fig. 1). Besides cytotoxicity, Aβ appears to impair the function of synapses. AβPP transgenic mice show cellular, biochemical, and electrophysiological evidence of synaptic dysfunction prior to degeneration of synapses and before Aβ deposition. Results are similar in studies with AD patients [14]. There are reports that this synaptic dysfunction is caused by soluble oligomeric Aβ. However, synaptic degeneration is common to several neurodegenerative diseases. Furthermore, synaptic impairment is also mediated by cytokines. The effect of cytokines depends on their concentration or other environmental factors. Interleukin 1β (IL-1β) is involved in hippocampal-dependent memory and long-term potentiation (LTP) under physiological conditions, whereas high levels of IL-1β impair memory and neural plasticity and induce synaptic depression. In light of these observations, an alternative interpretation for the synaptic effect of oligomeric Aβ is that early synaptic dysfunction in AD could also be the result of inflammation, secondary to aging or to the presence of preexisting pathology.

On the other hand, Aβ and proinflammatory cytokines have synergic effects [13]. Whereas the reactivity of glial cells exposed to Aβ derivatives is low, it is potentiated in the presence of proinflammatory molecules (Fig. 2a), resulting in higher production of NO and cytotoxicity [15]. The synergy between Aβ/AβPP and proinflammatory factors led us to postulate that Aβ/AβPP-induced cytotoxicity

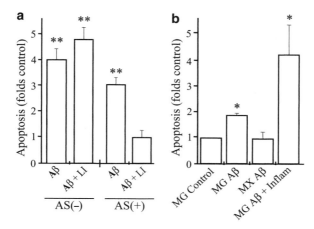

Fig. 1 Induction of cell death by Aβ . Aβ induces hippocampal cell death directly (**a**) or indirectly through the activation of glial cells (**b**). In (**a**), hippocampal cultures without (*AS−*) and with astrocytes (*AS+*) were unstimulated (control, not shown) or treated with 2 μM fibrillar Aβ$_{1-42}$ (Aβ) with or without 1 μg/ml LPS + 10 ng/ml IFNγ (LI) for 48 h. Neuronal cell death was evaluated by the TUNEL method with immunocytochemistry for β-Tubulin III as neuronal identity marker. Data from three independent experiments in triplicate is expressed as fold-number increase (mean ± SEM) compared to the control condition. $^*p < 0.01$, $^{**}p < 0.001$. In (**b**), indirect neurotoxicity was evaluated in hippocampal cultures exposed to conditioned media of microglial untreated (MG control) or exposed to Aβ (MG Aβ) with or without additional proinflammatory conditions (MG Aβ + Inflamm) at the indicated concentrations. Also mixed glial (microglia and astrocytes) cultures were exposed to Aβ (MX Aβ). Astrocytes show a protective effect on both direct and indirect Aβ neurotoxicity. Inflammatory conditions potentiate microglia-mediated toxicity, whereas it can also promote neuroprotection mediated by astrocytes

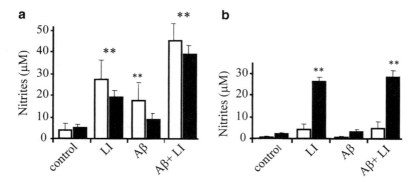

Fig. 2 Induction of nitrites secretion in glial and hippocampal cultures exposed to proinflammatory conditions. Microglial (□) or mixed glial cultures (■) in (**a**) and hippocampal (□) or hippocampal-microglia cocultures (■) in (**b**) were left unstimulated (control) or exposed to 2 μM Aβ$_{1-42}$ (Aβ) with or without 1 μg/ml LPS + 10 ng/ml IFNγ (LI). Data is expressed as mean ± SEM of three independent experiments in triplicate. $^{**}p < 0.001$. Astrocytes inhibit induction of NO production by microglia when exposed to Aβ or LI. However, inhibition is abolished after simultaneous treatment with Aβ and LI. Similarly, hippocampal cultures were poorly reactive to Aβ or LI, but NO production was greatly induced when hippocampal-microglia cocultures were exposed to proinflammatory mediators

only occurs under certain inflammatory conditions. Because such conditions also appear to influence the response of microglial cells to modulation by astrocytes, I propose that dysfunctional glial cells could be the centerpiece of AD pathogenesis [12]. Chronic inflammation results in the dysfunction of glial cells, reducing their scavenger function while potentiating their inflammatory-cytotoxic activation.

4 Inflammation in AD

The cellular bases for neuroinflammation are, among others, reactive microglia and astrocytes. Increased glial activation could be the result of aging in the absence of a pathological process, or secondary to trauma, infection, hypoxia-ischemia, etc. Glial activation is diverse depending on the basal functional state. Different outcomes (i.e., proliferation, phagocytosis, upregulation and release of cytokines or neurotrophic factors, and production of mediators such as NO and ROS) will be differentially regulated. Factors such as age, cell damage, metabolism, or oxidative stress are just some of the many factors that will influence an inflammatory response. Neuroinflammation requires the coordinated activity of several intracellular pathways. Coestimulation by various proinflammatory signals could be additive, show synergy or inhibition as manifestation of a regulatory cross talk, or saturation if a common upstream signaling component is involved.

Epidemiological evidence and experimental models of AD show that proinflammatory conditions promote the development of AD. Inflammation, microglial activation, and oxidative stress appear to precede the appearance of AD cytopathology [16]. Identification of susceptibility genes influencing the inflammatory process revealed that polymorphisms for IL-1α and IL-1β are associated with increased risk of early onset AD and the C allele of IL-6 associated with reduced IL-6 activity delays the onset of sporadic AD. Thus, inherited variations in inflammation mechanisms could influence AD pathogenesis. IL-1β, IL-6, transforming growth factor-β (TGF-β), and the inducible form of cyclooxygenase (COX-2) are elevated in the brain of AD patients. Likewise, glial cells surrounding amyloid plaques or exposed to Aβ in vitro secrete proinflammatory molecules such as tumor necrosis factor-α (TNF-α), IL-1β, monocyte chemotactic protein-1 (MCP-1), RANTES (CCL5), and eicosanoids. IL-1 expression is induced very early during plaque formation. In fact, IL-1α appears to be involved in the induction of AβPP expression after head trauma, suggesting a role for IL-1 in both amyloid plaque formation and neurodystrophic changes. However, research has failed to establish whether these changes are etiological or are secondary to the degenerative process. Nevertheless, clinicopathological studies and neuroimage show that inflammation and microglial activation precedes neuronal damage [17]. Also, case-control and population-based studies supported a roughly 50% reduction in AD risk after long-term use of nonsteroidal anti-inflammatory drugs (NSAIDs) [18]. The NSAID ibuprofen decreases cytokine-stimulated Aβ production and reduces neuritic plaques and inflammation in transgenic mice overexpressing AβPP [19]. NSAIDs have COX-independent effects, through various molecular mechanisms, including repression of β-secretase cleaving enzyme 1 (BACE 1), modulation of γ-secretase activity, and inhibition of signal transduction pathways involved in cytokine-mediated inflammation such as nuclear factor-κ B (NF-κB) and MAPKs. For example, decrease of Aβ production depends on changes of secretase-γ-cleavage without affecting COX activity.

Inflammation is involved in several pathophysiological mechanisms.

4.1   Inflammation and Aβ Load

There is evidence of several molecular mechanisms that indirectly influence amyloid production and probably contribute to neurodegeneration in AD. Brain injury produced by many different insults can induce Aβ deposition. Inflammation, likely involved in most age-associated neurodegenerative diseases, is linked to Aβ production and toxicity.

Besides their potential cytotoxic effects, TNF-α and IL-1β can promote AD by stimulating the synthesis of AβPP [20]. A similar effect has been described for TGF-β1 on astrocytes. Cytokines stimulate the processing of AβPP toward the generation of Aβ, including induction of β-secretase activity and inducing intracellular accumulation of Aβ. Furthermore, high concentrations of cytokines promote the formation of fibrils and aggregation of Aβ, inducing neuropathological changes similar to those found in AD brains [21].

4.2   Inflammation and Cytotoxicity

Proinflammatory molecules and NO synergize with hydrogen peroxide increasing neuronal sensitivity to oxidative injury. The release of NO both in culture [22] and in vivo is one of the mechanisms by which activated glial cells exert their cytotoxicity. McGeer and McGeer (1995) [23] proposed that it is not the accumulation of Aβ but rather the unregulated inflammatory response to it, responsible for neuronal damage in AD. A substantial part of our results strengthens that hypothesis and suggests a differential participation for microglia and astrocytes (Fig. 1). IL-1β and TNF-α potentiate cytotoxicity, inducing the expression of inducible nitric oxide synthase (iNOS) through activation of NF-κB, extracellular signal-regulated kinases (ERK), and jun N-terminal kinase (JNK) signal transduction pathways. However, IL-1 can exert beneficial effects, particularly when released in modest amounts. It promotes remyelinization and upregulation of growth factors needed for neuronal survival. IL-1β also increases TGF-β1 production and has a synergic effect with the neuroprotective activity of nerve growth factor (NGF), further promoting neuronal viability. Upregulation of TNF-α has been involved mostly with cell death. However, it also has a protective role in animal models for demyelizating diseases and traumatic brain injury.

4.3 Microglial Cell Activation: Inflammation Versus Aβ Phagocytosis

There is evidence to suggest that there is an association between inflammation and the phagocytic activity of glial cells. Whether intraperitoneal injection of AβPP transgenic mice with lipopolysaccharide (LPS) activates microglial cells and results in the transient clearance of Aβ deposits, at later times Aβ uptake is inhibited. In vitro assays also show that proinflammatory conditions increase Aβ uptake at early times while decreasing it at later times while glial cell damage rapidly evolves. If microglial cells become unable to clear Aβ under sustained inflammatory conditions, conditions favoring an increase in cytokines will result in Aβ accumulation over time, as well as the persistence of microglial cells activation. Those observations suggest that failure to adequately deal with the inciting stimulus (i.e., Aβ) in association with an inflammatory response results in the subsequent overactivation of microglia, which become detrimental, inducing the release of potentially toxic mediators.

4.4 Inflammatory Cytokines and Neuronal Function

Cytokines influence neuronal and synaptic function via diverse mechanisms, including regulation of neurotransmission, neurotransmitters receptors, and synaptic efficacy. Moderately high levels of IL-1 and TNF-α inhibits synapses, contributing to cognitive impairment. Cohort studies of healthy aging individuals showed that high levels of IL-6 correlates with lower cognitive functioning, also predicting subsequent cognitive impairment.

4.5 TGF-β1-Dependent Protective Mechanisms

TGF-β1 has prominent roles in tissue development, homeostasis, and repair. In the CNS, TGF-β1 has direct protective effects on neurons through the induction of cell survival-promoting proteins like Bcl-2 and proteins that participate in calcium homeostasis and modulation of MAPK, and indirect neuroprotective effect by preventing overactivation of microglial cells. It reduces Aβ-induced neurotoxicity and we have shown that TGF-β1 secreted by hippocampal cells modulates the production of NO and ROS by microglial cells [24] (see Sect. 5.4.3). Its expression is increased in the injured brain and in several pathological states including ischemia and neurodegenerative pathologies. Proinflammatory conditions induce production of TGF-β by astrocytes and hippocampal cells in vitro. TGF-β1 increases with age in the nervous system, apparently in response to glial activation. AD patients and AβPP transgenic mice show elevated levels of TGF-β1. However, Smad proteins, the major transduction pathway mediating its anti-inflammatory effects, show reduced levels and altered subcellular distribution. Changes on the signal transduction pathways of TGF-β1, especially if one of them is differentially affected, will have a profound effect on TGF-β1-mediated regulation.

5 Glial Cells and Neuroinflammation in AD

Both microglia and astrocytes produce several trophic factors and extracellular matrix molecules that are needed for the maintenance and function of neurons. Such functions are also meaningful in AD. Evidence suggests that glial cells activate in response to injury, illness, aging, or other stimuli. Factors secreted by microglial cells in response to injury induce activation of astrocytes. Activated astrocytes secrete growth factors that not only promote microglial growth and activation but also modulate their cytotoxicity. In consequence, glial activation can both impair and promote neuronal survival. An oversimplistic view is that microglia-derived factors are responsible for neurotoxicity whereas astrocytes secrete neuroprotective factors. However, astrocytes and microglia work cooperatively, exerting mutual regulatory activity. On the other hand, activation of microglia and astrocytes can be differentially affected by various stimuli. For example, hypoxia potentiates LPS induction of iNOS and TNF-α in microglia, whereas it is inhibitory for astrocytes. Aβ and inflammatory cytokines probably also activate glial cells through different mechanisms. Several signaling pathways are involved in microglial and astrocyte activation by Aβ, including tyrosine kinase-dependent pathways and NF-κB. Several of these pathways are common with those activated by inflammatory cytokines [25].

5.1 Microglial Cells

Microglial cells are resident macrophage cells and orchestrate the inflammatory response of the CNS. Aging is one of the factors modifying microglial activation and function [26]. Aged microglia express increased amounts of inflammatory cytokines, and because of their autocrine effect, a new steady state of activation is established, affecting microglial cell function. Microglia express more anti-inflammatory cytokines (TGF-β1 and IL-10) than proinflammatory cytokines [IL-1β, IL-6, IL-12, interferon-γ (IFN-γ) and TNF-α] when unstimulated. However, stimulation with a strong activator upregulates proinflammatory cytokines while anti-inflammatory cytokines are downregulated.

Most neurodegenerative diseases show increased numbers of microglia. Autopsy studies show microglial cell activation in AD patients and positron emission tomography reveals microglial activation in specific cortical regions at early stages of AD. It has been proposed that microglial cells associated with amyloid plaques are responsible for oxidative stress. However, we have observed that generation of ROS by microglia exposed to Aβ in vitro is minor. This apparent contradiction could depend on the fact that microglial cells in vivo are not activated solely by Aβ, but by a combination of proinflammatory stimuli. In agreement with that notion, an important activation is observed when glial cells are exposed simultaneously to Aβ plus inflammatory cytokines (Fig. 2).

There is also controversy whether activation of microglia is beneficial or harmful. Probably both effects are possible, depending on the specific activation pathways involved. Several of the modulators released by microglia are needed for the activation of protective mechanisms and induction of neurotrophic factors. Different activation patterns will generate different outcomes in terms of tissue damage and Aβ removal. Microglia can produce high amounts of ROS without becoming phagocytic and neutralize the offender, or vice versa. Ignoring the complexity of glial activation is an oversimplification that precludes the understanding of the neuroimmune response.

5.2   Astrocytes

Astrocytes are the structural and trophic support needed for neuronal housekeeping, synapse formation, and modulation. Astrocytes play an important role in inflammatory processes and have become a focal point in research because of their neuroprotective role following excitotoxicity, oxidative stress, ischemia, and Aβ toxicity. Mediators released by astrocytes trigger a dual response. They activate neighboring cells and amplify innate immune response; and they modify brain–blood barrier and attract circulating immune cells. The balance between inflammatory potentially tissue damaging and immunomodulatory functions is fundamental for controlled reaction to CNS insult.

In AD brains, astrocytes' activation is prominent around Aβ deposits. However, evidence is contradictory as to whether reactive astrocytes at sites of Aβ deposition have damaging or neuroprotective functions. Cytotoxic mediators produced by astrocytes, including proinflammatory mediators such as IL-1, MCP-1, RANTES, and TNF-α [27], have deleterious effects in AD and are involved in the pathogenesis of various neurodegenerative diseases. However, they also inhibit microglial cytotoxicity [13, 27] and can produce high levels of anti-inflammatory cytokines such as TGF-β1 (see Sect. 5.4.3) both in vitro and in vivo. Many of the neuroprotective effects depend initially on the inflammatory activation of glial cells, in agreement with the notion that inflammation also has beneficial roles. Impairment of these roles can unbalance their participation in favor of the deleterious effects. For example, because TGF-β1 production can be inhibited by TNF-α, it is possible that persistent inflammation could downregulate astrocytes' protective anti-inflammatory responses.

5.3   Modulation of Glial Activation: Glial Cells and Neurons

Microglial cytotoxic activity is modulated in various ways, including the induction of antioxidant enzymes and anti-inflammatory cytokines (IL-1Ra, IL-4, IL-10, and

TGF-β1). Their activity is also modulated by astrocytes and neurons. Feedback mechanisms through cross talk of brain cells restrain the amplitude and duration of glial activation. Neurons have an inhibitory role on glial activation by regulating the production of soluble factors and many proteins associated with reactive gliosis. Neurons and astrocytes modulate microglial cell reactivity. Microglial cells in pure cultures are more reactive than cocultures containing either astrocytes or neurons (Fig. 2). Active healthy neurons provide inhibitory factors including OX2, neurotrophins, and cytokines, which suppress the inflammatory response of microglia and astrocytes. In contrast, damaged neurons induce glial activation [28], producing inflammatory mediators, eicosanoids, C-reactive protein, amyloid protein, and complement factors.

Astrocytes remove toxic factors secreted by microglia and reduce microglial cell activation, enhance their phagocytic capability, and attenuate the production of ROS and TNF-α. We observed that astrocytes or soluble factors secreted by them abolish induction of NO by inflammatory conditions. Inhibition was also observed on the activation of the signal transduction pathways activated by those inflammatory conditions (Fig. 3). Activated astrocytes prevent Aβ-induced neurotoxicity in vitro [13], a protective effect that depends on diffusible factors. One of the modulator cytokines identified is TGF-β1 (*see* Sect. 5.4.3). Taken together, evidence suggests that astrocytes exert key functions in the modulation of neuroinflammation and have major regulatory effects on microglia and neurons exposed to Aβ. However, under strong inflammatory stimuli, astrocytes are unable to inhibit NO production by microglial cells [13, 15]. The duration of the exposure of astrocytes to proinflammatory stimuli is especially relevant for their activation and their modulation of microglial cell activation. Astrocytes exposed to proinflammatory conditions for 48 h or longer fail to modulate microglial cell activation, potentiating their neurotoxicity. These results suggest that persistent inflammatory conditions could impair regulatory functions normally performed by astrocytes, becoming an important pathogenic mechanism for neurodegenerative diseases.

5.4 Modulation of Glial Activation: Cytokines

Cytokines are constitutively expressed in glial and neuronal cells and exert actions in the normal brain, modulating various functions, including cell homeostasis, metabolism, synaptic function and plasticity, neural transmission, and complex behaviors. When upregulated in response to injury, infection, or various neurodegenerative and inflammatory conditions, primarily astrocytes and microglial, but also neurons, produce inflammatory mediators including chemokines, cytokines (e.g., TNF-α, IL-1, IL-6, IFN-γ, TGF-β), and prostaglandins. These mediators not only amplify the inflammatory response but are also protective, inducing glial cells to release neuroactive factors. The end result of glial activation depends on the profile of the secreted cytokines and other factors of the immediate environment.

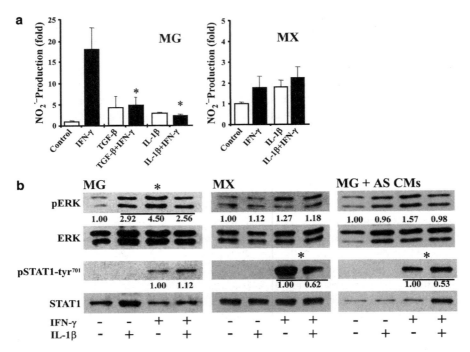

Fig. 3 Effect of IFN-γ, IL-1β, and TGF-β on NO secretion by microglial cell and mixed glial cultures. (**a**) Concentration of nitrites in the supernatant of microglial cell (*MG*) and mixed glial cultures (*MX*) nonstimulated (*control*) or incubated with 10 ng/ml IFN-γ, 250 pg/ml IL-1β, 1 ng/ml TGF-β, or combinations: IFN-γ/IL-1β and IFN-γ/TGF-β for 48 h. Both IL-1β and TGF-β decreased IFN-γ-induced NO secretion by microglial cells whereas IFN-γ-induced NO secretion in mixed glial cultures was nearly abolished. Values correspond to the mean ± SEM of three independent experiments in triplicate. $^*p < 0.05$ indicate conditions that were statistically different when compared with cultures exposed to IFN-γ. (**b**) Modulation of IFN-γ-induced activation of ERK and STAT1α pathways by IL-β at 30 min. MG, MX, or microglia cultured with astrocytes' conditioned media (*MG + AS CMs*) were exposed to 10 ng/ml IFN-γ, 250 pg/ml IL-1β, or the combination IFN-γ/IL-1β. IL-1β inhibited ERK phosphorylation induced by INF-γ in microglial cells, whereas it decreased STAT1α phosphorylation induced by INF-γ in mixed glial cultures and microglial cells exposed to astrocytes' conditioned media. Numbers indicate the ratio of the activated and total fractions of ERK or STAT signal transduction pathways. * indicates conditions associated to statistically significant differences

5.4.1 IL-1β Is Expressed by Neurons and Glial Cells

IL-1β increases during aging, associated with the increased reactivity of glial cells. It modulates inflammation and multiple cell functions in basal conditions. IL-1β can either contribute to or limit neuronal damage or death. Its beneficial effects are observed particularly when released in modest amounts [29]. It promotes remyelinization and the upregulation of growth factors. IL-1β also increases TGF-β1 and has a synergic effect with NGF. Activation of the IL-1 receptor, IL-1RI, leads to the translocation of NF-κB to the nucleus and the activation of the mitogen-activated protein kinase (MAPK) pathways ERK, p38 MAPK, and JNK1 (*see* Sect. 5.5).

5.4.2 TNF-α in Health and Disease

TNF-α is known to be expressed in pathological conditions and has been implicated in the pathogenesis of several neurological diseases. However, its expression in the healthy brain is controversial. TNF-α modulates neuronal function, affecting synaptic activity and the release of neurotransmitters. Depending on the physiological state of its target cell, TNF-α exerts pleiotropic effects on neurons, including toxic and protective effects. When upregulated, TNF-α plays a key role as mediator of tissue injury and inflammation.

5.4.3 TGF-β1 Has Prominent Roles in Tissue Development, Homeostasis, and Repair

The concentration of TGF-β1 is low in normal brain tissue, whereas its expression is increased in activated glial cells in the injured brain and in several pathological states. Its concentration in the CNS increases with aging, apparently in response to glial activation. TGF-β1 is synthesized in the brain in response to insults such as ischemia and various neurodegenerative pathologies and it has been identified as one of the factors involved in neuroprotection secreted by astrocytes [30]. It modulates microglial activation inhibiting production of IL-1 and TNF-α, expression of the major histocompatibility complex class II (MHC-class II) and Fas glycoprotein, NOS induction, and release of NO and O_2^- production [24]. Treatment with exogenous TGF-β1 has a protective effect on Aβ-induced neurotoxicity. Hippocampal cells and activated astrocytes secrete TGF-β1 both in vivo and in vitro. TGF-β1 is upregulated in hippocampal neurons in ischemia and in astrocyte cultures exposed to IFN-γ, LPS, or TNF-α.

5.5 Modulation of Glial Activation: Cytokines and Their Signal Transduction Pathways

Cytokines activate a number of signaling mechanisms, including major effectors pathways and signal pathways serving modulator functions. Activation by most if not all cytokines leads to the activation of MAPK ERK, P38, and JNK, as modulator pathways. MAPK regulation in neuroinflammation allows for the integration of specific signaling pathways to yield unique output responses. It could be involved in both the magnitude and the temporal pattern of the response.

The combinatorial complexity of signaling may account for the fact that depending on a specific background determined by intrinsic cell characteristics and environmental factors, proinflammatory stimuli can result in limited inflammation and repair or cell damage and degenerative changes. For example, IL-1β plus TNF-α is needed in order to induce the expression of proinflammatory cytokines and iNOS through the activation of NF-κB and ERK. Although IL-1β or TNF-α is capable of activating NF-κB by itself, iNOS induction requires additional transcription factors activation modulated by MAPKs. Furthermore, increased expression and amyloidogenic processing of APP induced by inflammation appears to require activation of ERK and JNK in order to be generated.

The main signal pathway induced by IFN-γ is the signal-transducer and activator of transcription-1 (STAT1) in addition of ERK and P38 MAPKs. STAT-1 is the main transcription factor involved in the induction of iNOS by IFN-γ. It is activated by a JAK-dependent phosphorylation of a tyrosine residue (pSTAT1[tyr]). ERK plays an important modulator role because full activation of STAT-1 depends on a second phosphorylation at a serine residue (pSTAT1[ser]) mediated by ERK.

TGF-β1 exerts its function mainly through receptor-activated Smad signaling. It also has neuroprotective effects through the activation of PI3 kinase/Akt and ERK pathways; and the modulation of glial cell activation is mediated by regulation of ERK and P38. TGF-β1-dependent activation of MAPKs is transient and renders microglial cells refractory to further activation by proinflammatory cytokines. The inhibition by TGF-β1 is produced by a strong expression of MAPK phosphatase type 1 (MKP-1), which is persistently upregulated under proinflammatory conditions.

Evidence supports the existence of a complex regulatory interaction between TGF-β1 and IFN-γ. IFN-γ null mice express high amounts of TGF-β1 and show an increased activation of Smad signaling pathway. In contrast, TGF-β1 null mice have a high level of IFN-γ and overexpress STAT-1, iNOS, and NO production. Overactivation of STAT-1 in TGF-β1 null mice supports the notion that TGF-β1 is an essential immune-regulator for the control of inflammatory events. In glial cells, TGF-β1 reduces IFN-γ-induced pERK, STAT-1 phosphorylated at serine[727] (pSTAT-1[ser]) and tyrosine[701] (pSTAT-1[tyr]), and total STAT-1.

Under normal conditions, TGF-β1 modulates microglial cells activation and abolishes neurotoxicity. Its induction by proinflammatory molecules would naturally limit the temporal and spatial extent of the inflammatory response. TGF-β1 modulator effects involve activation of the Smad3 pathway and the activation of MAPK-ERK pathways. At least one of these pathways, Smad3, is downregulated in AD patients. Dynamic regulation of Smad, PI3K, and MAPK pathways, which are associated to TGF-β activity as well as to other inflammatory cytokines, can be key elements for cellular integrity and handling of inflammation. Disappearance or reduction of Smad3, a major effectors pathway for anti-inflammatory modulation will modify the regulation and feedback signals resulting from inflammation, impairing normal regulatory mechanisms.

We have shown that ERK is involved in the modulation of microglial response by TGF-β1 and IL-1β, both decreasing NO production. Modulation of oxidative molecules involved ERK and p38 MAPKs transduction pathways. IL-1β and TGF-β1 inhibit IFN-γ-induced ERK phosphorylation with different kinetics; IL-1β inhibition of ERK phosphorylation after 30 min of activation with IFN-γ is short-lived. In contrast, TGF-β1 inhibits ERK phosphorylation only after several hours of activation and inhibition persists for a long time (Fig. 4). IL-1β, which is induced

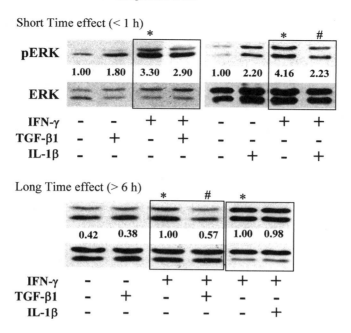

Fig. 4 Time dependence of the modulation of IFN-γ-induced activation of ERK by TGF-β1 and IL-1β in glial cells. Western blots of microglial cell cultures treated with 10 ng/ml IFN-γ in the presence or absence of 1 ng/ml TGF-β1 or 250 pg/ml IL-1β for short times (less than 60 min) and long times (more than 6 h). IL-1β decreased by 50% pERKs induced by IFN-γ at short times of exposure, whereas TGF-β1 decreased activation of ERK only after exposures of 6 h or longer. Below each blot, the mean of the densitometry analysis of three to five independent experiments is shown. $^*p < 0.02$ between the cultures treated with IFN-γ and the control cultures, $^\#p < 0.03$ between cultures treated with IFN-γ and those cotreated with IFN-γ and TGF-β1 or IL-1β

early after injury, could mediate the autoregulation of microglial cells activation. If microglia are unable to resolve the activating condition through IL-1β-dependent modulation, TGF-β1 pathways would be activated at later stages to further attempt modulation of the inflammatory process. Those results led us to propose that impairment of the modulator effect of TGF-β1, as observed when specific signal pathways are downregulated, promotes cytotoxicity.

6 Glial Dysregulation Hypothesis in AD

Regulation of glial activation appears to be impaired under sustained proinflammatory conditions. Aging associates with a proinflammatory status in the CNS inducing functional changes on microglia and astrocytes. These changes include a shift in basal cell reactivity and changes on modulation mechanisms (i.e., inhibition of TGF-β Smad pathway) that can lead to regulatory impairment and promote neuro-degenerative processes. We observed that whereas basal microglial reaction to Aβ is mild and downregulated by astrocytes, regulation is lost under proinflammatory conditions. These conditions are also associated with decreased uptake of Aβ. I propose that AD is the result of dysfunctional microglia [12], which, by not responding to normal regulatory feedback mechanisms and/or having an impaired ability to clear Aβ, loose their ability to handle potentially toxic compounds and become cytotoxic because of persistent inflammatory activation.

Inflammatory conditions potentiate the response of glial cells to other stimuli and can result in an increased reactivity when exposed to a condition that normally would fail to induce their activation. That appears to be the case for Aβ, a compound that is a not potent activator by itself [13, 15]. Abnormal glial cell activation can also impair the capacity of glia to uptake and degrade Aβ. The persistence of Aβ can establish a vicious circle, further potentiating an inflammatory reaction. In time, chronically or multiple-event activated microglia and astrocytes would become neurotoxic by the release of inflammatory cytokines, proteolytic enzymes, complement factors, and reactive intermediaries.

7 Relevance of Glial Cells in AD Therapy

AD disease-modifying therapies presently under evaluation could be loosely grouped in three categories, anti-amyloid, neuroprotective, and neurorestorative. Of the multiple mechanisms investigated, few are related to inflammatory mechanisms, and even more scarce are those specifically targeting glial cell regulation (in contrast to general glial suppression). The participation of neuroinflammatory mechanisms in the genesis or progression of neurodegenerative diseases has led to investigation of the therapeutic effect of anti-inflammatory and antioxidant treatments. Whereas epidemiological studies suggest that the risk of suffering AD is reduced in patients treated with NSAIDs; most trials for anti-inflammatory agents report no significant beneficial effect. The results are similar for randomized controlled clinical trials for antioxidants, such as vitamin E, also showing only marginal benefits.

Other therapeutic approaches are targeted toward microglial cell inhibition [31, 32]. However, there is clear evidence that microglial activity is needed for normal function and Aβ clearance. When their function is reduced or abolished, it may result in extensive damage and even worsening of AD pathology [33]. A better understanding of the regulation of glial cell activation and their participation in Aβ clearance should be a key element on any AD therapeutic approach, targeting neuroinflammation and plaque removal. A successful treatment should promote repair and phagocytic functions of microglia while inhibiting cytotoxic activation.

8 Conclusion

I propose that glial dysfunction associated with inflammatory or oxidative stress (i.e., as in aging) are the principal culprits of sporadic AD, whereas increased production/levels and deposition of Aβ are secondary to glial dysfunction. It is well reported that deposition of Aβ can occur after oxidative stress, secondary to neuronal stress and in inflammatory conditions. These observations suggest that the increased generation of Aβ is not the primary cause of sporadic AD, but part of the CNS response to injury, oxidative stress, or inflammatory insult.

Acknowledgments I thank Dr. Jaime Eugenín for his longstanding support and his critical reading of the manuscript. I gratefully acknowledge technical support of G. Ramírez and support by grant 1040831 from FONDECYT to RvB.

References

1. Duyckaerts C, Colle MA, Dessi F, Crignon Y, Piette F, Hauw JJ (1998) The progression of the lesions in Alzheimer disease - insights from a prospective clinicopathological study. J Neural Trans Suppl 53: 119–126
2. de la Torre JC (2002) Alzheimer's disease as a vascular disorder: nosological evidence. Stroke 33: 1152–1162
3. Letiembre M, Hao W, Liu Y et al (2007) Innate immune receptor expression in normal brain aging. Neuroscience 146: 248–254
4. Fassbender K, Walter S, Kuhl S et al (2004) The LPS receptor (CD14) links innate immunity with Alzheimer's disease. FASEB J 18:203–205
5. Streit WJ, Sammons NW, Kuhns AJ, Sparks DL (2004) Dystrophic microglia in the aging human brain. Glia 45: 208–212
6. Alarcón R, Fuenzalida C, Santibañez M, von Bernhardi R (2005) Expression of scavenger receptors in glial cells: comparing the adhesion of astrocytes and microglia from neonatal rats to surface-bound β-amyloid. J Biol Chem 280: 30406–30415
7. Wegiel j, Imaki H, Wang KC, Wronska A, Osuchowski M, Rubenstein R (2003)) Origin and turnover of microglial cells in fibrillar plaques of APPsw transgenic mice. Acta Neuropathol (Berlin) 105: 393–402
8. Simard AR, Soulet D, Gowing G, Julien JP, Rivest S (2006) Bone marrow-derived microglia play a critical role in restricting senile plaque formation in Alzheimer's disease. Neuron 49: 489–502
9. Husemann J, Loike JD, Anankov R, Febbraio M, Silverstein SC (2002) Scavenger receptors in neurobiology and neuropathology: their role on microglia and other cells of the nervous system. Glia 40: 195–205
10. Mantovani A, Sica A, Locati M (2007) New vistas on macrophage differentiation and activation. Eur J Immunol 37: 14–16
11. Wyss-Coray T, Loike JD, Brionne TC, et al (2003) Adult mouse astrocytes degrade amyloid-β in vitro and in situ. Nat Med 9: 453–457
12. von Bernhardi R (2007) Glial cell dysregulation: a new perspective on Alzheimer disease. Neurotoxicity Res 12: 1–18
13. von Bernhardi R, Eugenín J (2004) Microglia - astrocyte interaction in Alzheimer's disease: modulation of cell reactivity to Aβ. Brain Res 1025: 186–193
14. Selkoe DJ (2002) Alzheimer's disease is a synaptic failure. Science 298: 789–791
15. von Bernhardi R, Ramírez G, Toro R, Eugenín J (2007) Pro-inflammatory conditions promote neuronal damage mediated by Amyloid Precursor Protein and degradation by microglial cells in culture. Neurobiol Dis 26: 153–164
16. Zhu X, Raina AK, Perry G, Smith MA (2004) Alzheimer's disease: the two-hit hypothesis. Lancet Neurol 3: 219–226
17. Eikelenboom P, van Gool WA (2004) Neuroinflammatory perspectives on the two faces of Alzheimer's disease. J Neural Transm 111: 281–294

18. Etminan M, Gill S, Samii A (2003) Effect of non-steroidal anti-inflammatory drugs on risk of Alzheimer's disease: systematic review and meta-analysis of observational studies. BMJ 327: 128–131

19. Lim GP, Yang F, Chu T et al (2000) Ibuprofen suppresses plaque pathology and inflammation in a mouse model of Alzheimer's disease. J Neurosci 20: 5709–5714

20. Rogers JT, Leiter LM, McPhee J et al (1999) Translation of the Alzheimer amyloid precursor protein mRNA is up-regulated by interleukin-1 through 5 -untranslated region sequences. J Biol Chem 274: 6421–6431

21. Sheng JG, Bora SH, Xu G, Borchelt DR, Price DL, Koliatos VE (2003) Lipopolysaccharide-induced-neuroinflammation increases intracellular accumulation of amyloid precursor protein and amyloid beta peptide in APPswe transgenic mice. Neurobiol Dis 14: 133–145

22. Bal-Price A, Brown C (2001) Inflammatory neurodegeneration mediated by nitric oxide from activated glia-inhibiting neuronal respiration, causing glutamate release and excitotoxicity. J Neurosci 17: 6480–6491

23. McGeer PL, McGeer ED (1995) The inflammatory response system of brain: implications for therapy of Alzheimer and other neurodegenerative diseases. Brain Res Rev 21: 195–218

24. Herrera-Molina R, von Bernhardi R (2005) Transforming growth factor-α1 produced by hippocampal cells modulates glial reactivity in culture. Neurobiol Dis 19: 229–236

25. Nguyen MD, Julien J-P, Rivest S (2002) Innate immunity: the missing link in neuroprotection and neurodegeneration? Nat Rev Neurosci 3: 216–227

26. Sierra A, Gottfried-Blackmore AC, McEwen BS, Bulloch K (2007) Microglia derived from aging mice exhibit an altered inflammatory profile. Glia 55: 412–424

27. Smits HA, van Beelen AJ, de Vos NM et al (2001) Activation of human macrophages by amyloid-beta is attenuated by astrocytes. J Immunol 166: 6869–6876

28. Sudo S, Tanaka J, Toku K et al (1998) Neurons induce the activation of microglial cell in vitro. Exp Neurol 154: 499–510

29. Basu A, Krady J, Levinson S (2004) Interleukin-1: a master regulator of neuroinflammation. J Neurosci Res 78: 151–156

30. Dhandapani KM, Hadman M, De Sevilla L, Wade MF, Mahesh VB, Brann DW (2003) Astrocyte protection of neurons. Role of transforming growth factor-β signaling via c-jun-Ap-1 protective pathway. J Biol Chem 278: 43329–43339

31. McCarty MF (2006) Down-regulation of microglial activation may represent a practical strategy for combating neurodegenerative disorders. Med Hypotheses 67: 251–269

32. Garcia-Alloza M, Ferrara BJ, Dodwell SA, Hickey GA, Hy-man BT, Bacskai BJ (2007) A limited role for microglia in antibody mediated plaque clearance in APP mice. Neurobiol Dis doi:10.1016/j.nbd.2007.07.019

33. El Khuory J, Toft M, Hickman SE et al (2007) Ccr2 deficiency impairs microglial accumulation and accelerates progression of Alzheimer-like disease. Nat Med 13: 432–438

Cerebrovascular Pathology and AD

Color Plates

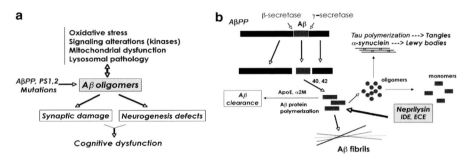

Chapter 1, Fig. 1 Mechanisms of Aβ toxicity and clearance. (**a**) Accumulation of Aβ oligomers might be involved in promoting synapse damage and neurogenesis defects. (**b**) Aβ-degrading enzymes such as neprilysin (*NEP*), insulin-degrading enzyme (*IDE*), and endothelin-converting enzyme (*ECE*) play a central role in Aβ clearance

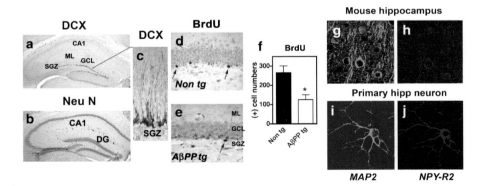

Chapter 1, Fig. 2 Neurogenesis in the hippocampus in AβPP tg mice and neuropeptide (*NPY*)-R expression. (**a–c**) Doublecortin (*DCX*)-positive neuronal precursor cells (*NPC*) in the hippocampus subgranular zone (*SGZ*). (**d and e**) In the mThy1-AβPP tg mice, the numbers of NPC in the SGZ are reduced. (**g–j**) NPY-R2 is colocalized in the neuronal marker MAP2 in the mature hippocampus and in primary hippocampal cultures (*n* = 12 mice per group)

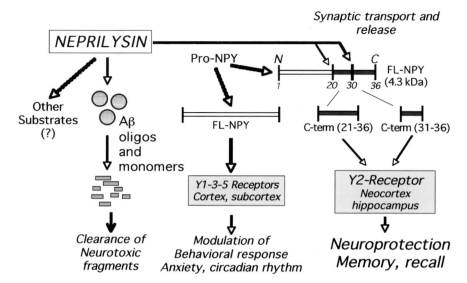

Chapter 1, Fig. 3 Metabolic processing of neuropeptide Y (*NPY*) in the CNS and neuroprotection. Neprilysin (*NEP*) might ameliorate the neurodegenerative pathology in Alzheimer's disease (*AD*) by reducing Aβ and by generating protective CTF-NPY that may bind the Y2 receptor

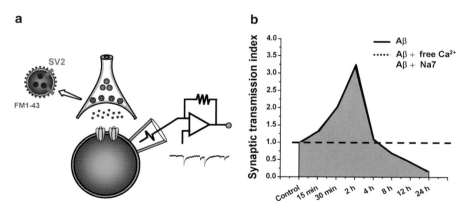

Chapter 2, Fig. 1 Effects of $Aβ_{1-40}$ oligomers on synaptic activity of hippocampal neurons. **a** The scheme illustrates pre- and postsynaptic components of a central synapse. The vesicles are released in a calcium-dependent fashion. Presynaptic activity can be determined by the presence of vesicular proteins (SV2) or by the staining of synaptic vesicles with fluorescent probes such as FM1–43 (*red dots*). The postsynaptic membrane currents associated to the vesicular release and postsynaptic receptor density can be analyzed using the patch clamp technique. **b** Time-dependent biphasic effect of 500 nM Aβ oligomers on synaptic transmission in hippocampal neurons. The effects of Aβ oligomers were blocked by lowering extracellular calcium or by adding Na7 (*broken line*). The values were obtained from three independent experiments

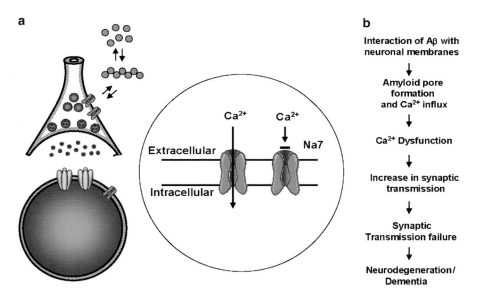

Chapter 2, Fig. 3 Proposed hypothesis that explains the effect of Aβ oligomers on synaptic transmission. **a** Aβ oligomers bind to the membrane inducing the formation of pores in the pre- and postsynaptic membranes. The pores allow calcium to enter the cell and this event can be blocked by Na7. **b** Proposed series of events that leads to AD. Its initiation depends on oligomerization and membrane perturbations that lead to a calcium dysfunction and alterations in synaptic transmission

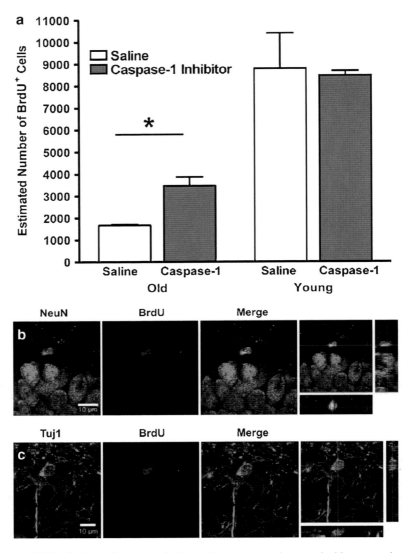

Chapter 3, Fig. 3 Age-related neuroinflammation causes a decrease in hippocampal neurogenesis. **a** Aging results in a significant decrease in neurogenesis, as was seen by the decrease in BrdU-postive cells in the subgranular zone/granule cell layer (*SGZ/GCL*) of aged fisher 344 rats compare to young adult rats. The age-related decrease in neurogenesis was attenuated by 28 days of intracerebroventricular infusion of Ac-YVAD-CMK, a capase-1 inhibitor ($^*P < 0.05$). Caspase-1 inhibition blocks the production of pro-IL-1β to the active form of interleukin (*IL*)-1β. The results demonstrate that at least part of the decrease in hippocampus neurogenesis with age is due to elevated cytokine levels. Representative confocal micrograph of (**b**) BrdU-labeled cells (red) and NeuN[+] (green) and (**c**) BrdU[+] (red) and TUJ1[+] (green), confirms that the increase in BrdU[+] cells did represent an increase in neurogenesis

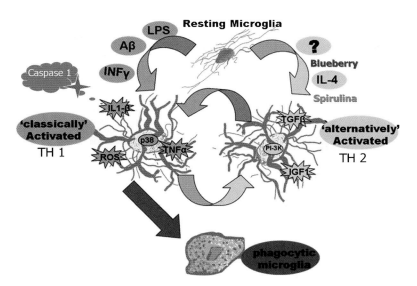

Chapter 3, Fig. 4 Microglia: protective or harmful? Microglia are normally in a resting state in which they are actively surveying the microenvironment of the brain. The microglia are resting in the sense that they are not performing effector functions such as producing inflammatory mediators like interleukin (*IL*)-1β and tumor necrosis factor-α (*TNFα*). When microglia are producing inflammatory mediators, the microglia would be considered in a "classically activated state" or "TH1" state. Microglia can also become "alternatively activated" in such a way that they produce growth factors, such as IGF-1 and TGFβ. The "alternatively activated" microglia can support tissue remodeling and repair. Beyond releasing signaling molecules, microglia also have an important role in phagocytosis. The role of microglia, as protective or harmful, depends upon the ability of the microglia to switch from the different activation states at the appropriate time. Understanding how and when to turn microglia "on" or "off" is an important future direction of research. This is especially the case with aging where microglia are most needed to remodel and repair and to remove damaged cells and misfolded proteins. With age, microglia may lose the ability to perform these important effector functions making the aged brain more susceptible to injury and insult

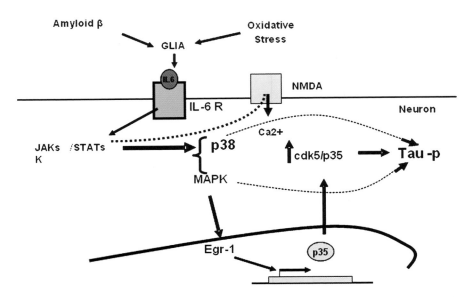

Chapter 5, Fig. 1 Schematic representation of the effects of Aβ and brain oxidative stress signals (redox iron, oxygen, and nitrogen free radicals) on glial cells, the resulting release of IL-6 proinflammatory cytokine (could be also IL-1, TNFα), and activation of neuronal receptors and signaling through JAKs/STAT system. This signaling cascade activates MAPK and p38. Phosphorylated active MAPK, via the translation factor Erg-1, activates p35 gene, among others, increasing its neuronal expression, in the overactivation of cdk5 and the subsequent anomalous tau hyperphosphorylations (Representation generated from data of Quintanilla et al. [59] and Orellana et al. [64, 65])

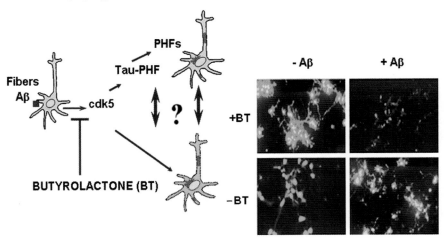

Chapter 5, Fig. 2 Schematic representation showing the effects of inhibition of the protein kinase cdk5 by butyrolactone I (*BT*) on hippocampal cells in primary cultures. cdk5 inhibition protected hippocampal cells against the Aβ-induced neurodegeneration and neuronal death. The inmunofluorescence photomicrographs correspond to analysis of cell viability in the primary cultures of hippocampal cells derived from rat 18-day embryos. Studies indicate that cdk5 inhibitors such as BT (and roscovitine) protect neuronal cells against the neurotoxic effects of 10 μM soluble Aβ oligomers (Representation generated on the basis of experimental data from Alvarez et al. [16, 17] and Muñoz et al. [68])

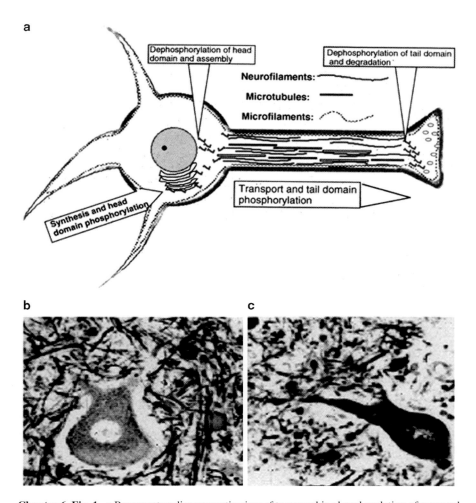

Chapter 6, Fig. 1   a Represents a diagrammatic view of topographic phosphorylation of neuronal cytoskeletal proteins, biosynthesis in the cell body, assembly of cytoskeletal proteins in axon-hillock region, phosphorylation and transport in the axonal compartment, and finally dephosphorylation and degredation at the nerve terminals. **b** Human cervical spinal cord neuron (physiology), no phosphorylation occurs in the cell body; however, neurites are selectively phosphorylated (SMI 31 staining). **c** Human cervical spinal cord neuron from ALS patient (pathology). Aggregates of aberrantly hyperphosphorylated deposit cytoskeletal proteins in the cell body

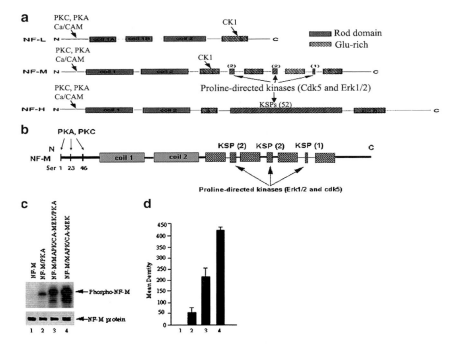

Chapter 6, Fig. 2 a Domain structures of NF-subunit proteins, NF-L (low molecular weight), NF-M (middle molecular weight), NF-H (high molecular weight), their phosphorylation domains by identified kinases, PKA (cAMP-dependent protein kinase), PKC (protein kinase C), Ca^{2+}/CAM (calcium–calmodulin-dependent kinase), proline-directed Ser/Thr kinases (Cdk5, Erk1/2, GSK3β), casein kinase I (CK I), and casein kinase II (CK II). **b** Phosphorylated residues in the head domain of NF-M by PKA. (**c**) Phosphorylation of bacterially expressed and purified NF-M in vitro. *Lane 1:* No kinase added, only NF-M, *Lane 2:* NF-M phosphorylated by PKA, *Lane 3:* NF-M first phosphorylated by PKA, then Erk1/2, and *Lane 4:* NF-M phosphorylated by Erk1/2. *Top panel* is autoradiograph and *bottom* is Commassie stain. **d** represents the quantitation of data shown in **c**

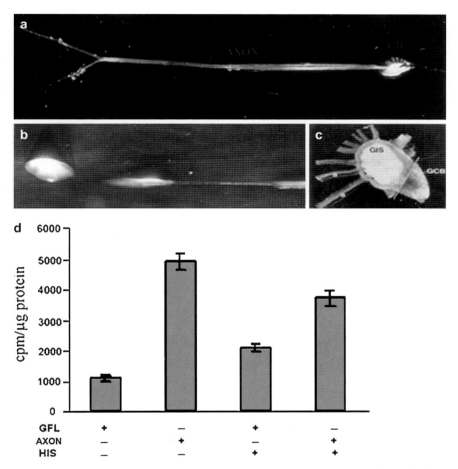

Chapter 6, Fig. 3 **a** Squid giant axon containing ganglion cell body. **b** Squeezed out axioplasm. **c** Ganglion cell body bag. **d** Endogenous and exogenous phosphorylation activity of axon and cell body preparations, endogenous (no histone H1) and exogenous (presence of histone H1)

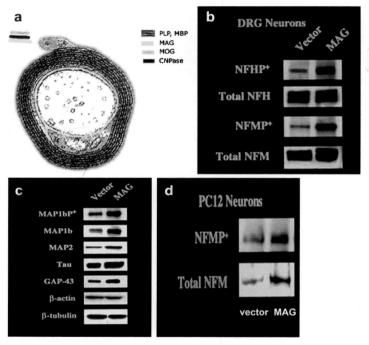

Chapter 6, Fig. 4 **a** Cross section of myelinated axon. Different layers represent identified myelin proteins. **b** Dorsal root ganglia isolated from rat were cocultured with COS cell containing empty vector (*left lanes*) and MAG stably transfected (*right lanes*), expression of NF-M/H was analyzed using phospho- and total NF-M/H-specific antibodies. **c** Expression of MAP1b, MAP2, Tau, GAD-43, β-actin, and tubulin was analyzed using respective specific antibodies to corresponding molecules. **d** PC-12 cells were cocultured with COS cell and transfected with vector alone (*left lanes*) and MAG-containing vector (*right lanes*) expression of phosphorylated and total NF-M was analyzed

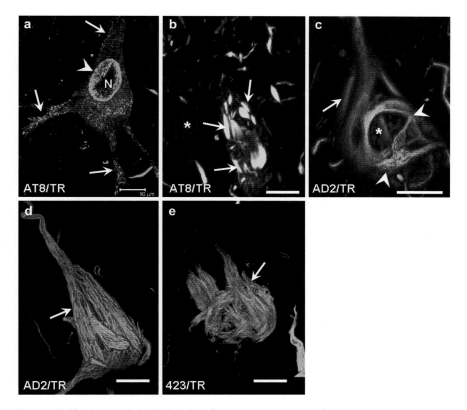

Chapter 7, Fig. 1 Morphological model of neuronal degeneration from the pretangle state to the formation of intracellular and extracellular tangles. **a** A pretangle stage neuronal cell. Stage 1. is characterized by diffuse granular deposits throughout the perinuclear area (*arrowhead*) and proximal processes, which are undetected by thiazin red (*TR*). **b** Stage 2. corresponds to the presence of bead-like structures (paired helical filaments, PHF, nucleation sites, *arrows*), some of which are detected by TR. **c** Stage 3. is characterized by a cell having long bundles of PHFs (TR positive) covering up the nuclear and lipofuscine areas (*arrow*), the networking processing of PHF-assembled tau (*arrow*), and the putative fusion of bundles (*arrowheads*). **d** Stage 4. Intracellular neurofibrillary tangles (*NFT*) (*arrow*) having the typical flame-like appearance. **e** Stage 5. Extracellular tangles (*arrow*) whose appearance is modified by the extracellular proteolytic process. These ghost tangles only contain the PHF core, which is identified by mAb 423 and remains detected by TR [29]. Double labeling with antibodies AT8 (**a, b**), AD2 (**c, d**), and 423 (**e**) and TR. Bar = 10 μm

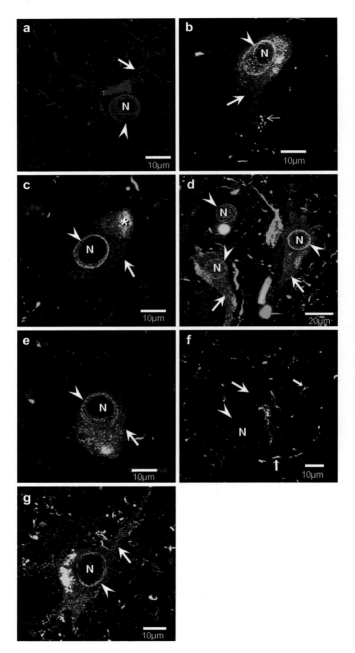

Chapter 7, Fig. 2 a, b Double immunolabeling with mAb TG-3 (*green channel*) and pT231 polyclonal antibody (*red channel*). (*For complete caption refer page 84*)

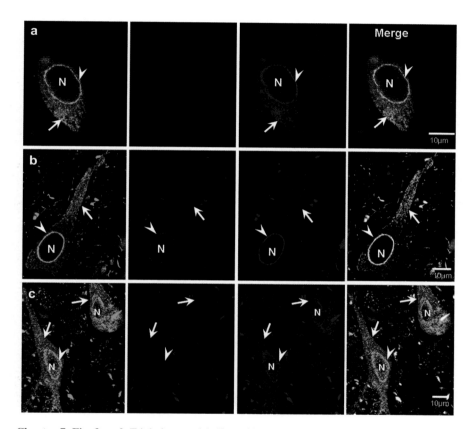

Chapter 7, Fig. 3 a, b Triple immunolabeling with antibodies pT231 (*green channel*), Tau-C3 (*red channel*), and TG-3 (*blue channel*). Triple immunolabeling with mAbs Tau-C3 (*red channel*) and TG-3 (*blue channel*) in pretangle cells. Cells illustrated in **a, b** are detected by both pT231 and TG-3; however, Tau-C3 immunoreactivity is present in **b** but absent in **a**. Colocation between pT231 and TG-3 is more evident in the perinuclear area (*arrowhead*). **c** Triple labeling with Alz-50 (*green channel*) and the C- and N-ends of tau protein (T46, *red channel* and M19G, *blue channel*) Alz-50 immunoreactivity is present in pretangle cells (*arrows*) and perinuclear mAb Alz-50 found a double labeling between Alz-50 and N-terminus epitopes (*blue channel*) but not between Alz-50 and the C-terminus epitope (*red channel*). *N* nucleus

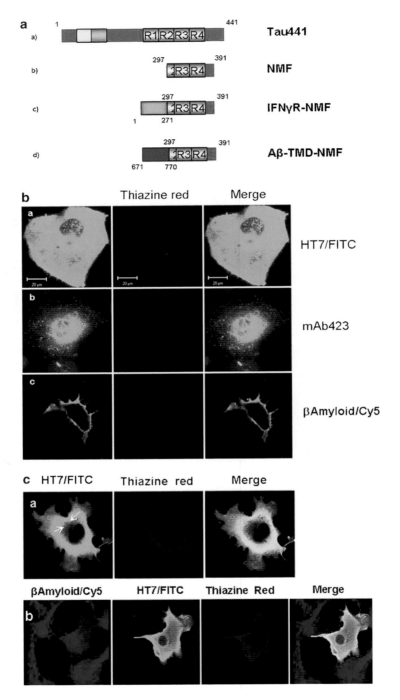

Chapter 8, Fig. 1 PHF-core anchored to plasma membrane induce the presence of β-sheet structures in COS7 cells culture. **a** Schematic representation of constructs used to transfect COS7 cell culture or transduce primary cultures of neural precursor cells: (a) human full-length (isoform 441)

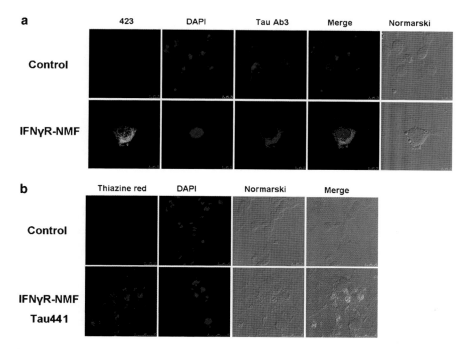

Chapter 8, Fig. 2 Generation of β-sheet structures in neural precursor cell (*NPC*) induced by IFNγR-NMF. **a** Immunofluorescence staining of endogeneous tau protein (antibody tau Ab3) showed a partial colocalization with IFNγR-NMF expression (mAb423). **b** Cells that express IFNγR-NMF and tau441 showed positive staining of β-sheet structures by TR

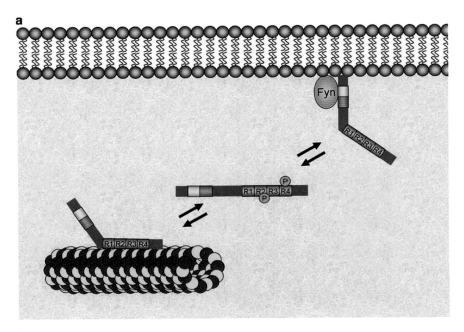

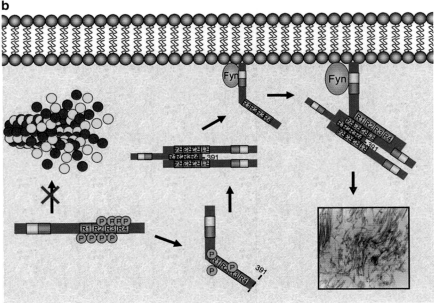

Chapter 8, Fig. 3 Hypothetical model of PHF assembly. See text for details

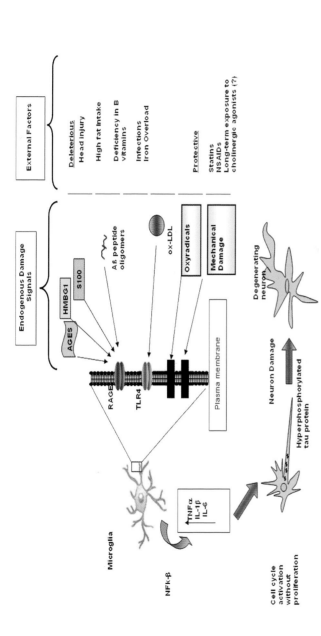

Chapter 11, Fig. 1 Schematic representation of the hypothetical roles of endogenous danger/alarm signals built into the innate immune system at the early stages of the pathogenesis of Alzheimer's disease (*AD*). Danger signals such as advanced glycation end products (*AGES*), high mobility box group 1 (*HMBG1*), S-100 proteins, and Aβ peptide oligomers (but not β-pleated fibrillar aggregates) activate microglia through the AGES receptor (RAGE; shown here as a transmembrane protein in a cartoon of a lipid bilayer). Separately, oxidized low-density lipoproteins (*oxLDL*) activate toll-like receptors (*TLRs*), particularly TLR4. Additional danger signals, such as trauma and oxyradical damage, possibly acting on separate receptors (*black boxes* inserted in the membrane) as well as by inducing the production of additional Aβ peptide oligomers, AGES, and S-100 protein contribute to this process. Separately and in various combinations (as it may apply to different individuals), these danger signals would trigger innate immune system alarm mechanisms resulting in long-term excessive levels of tumor necrosis factor-α (*TNFα*), interleukin 1-β (*IL1-β*), and IL-6. These signals would then mediate neuronal damage directly, reflected in alterations such as tau protein hyperphosphorylation and paired helical filaments formation, which eventually result in neuronal degeneration, interstitial Aβ peptide aggregation, and its deposition in a β-pleated configuration, and progressively more severe clinical manifestations of cognitive and behavioral decline unrelated to the amyloid aggregation process (Reproduced with permission from Fermandez et al., 2008)

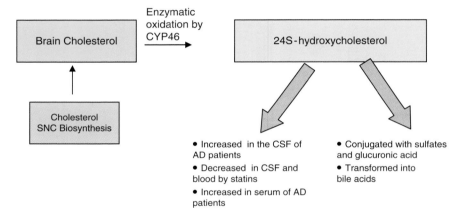

Chapter 11, Fig. 2 Brain cholesterol is mostly synthesized in the central nervous system. 24S-Hydroxycholesterol is a metabolite produced via enzymatic oxidation of brain cholesterol and is eliminated as sulfatides and glucuronides or transformed into bile acids. Blood levels of this cholesterol metabolite are normally decreased in elderly and abnormally increased in the serum and cerebrospinal fluid (*CSF*) of Alzheimer's disease (*AD*) patients

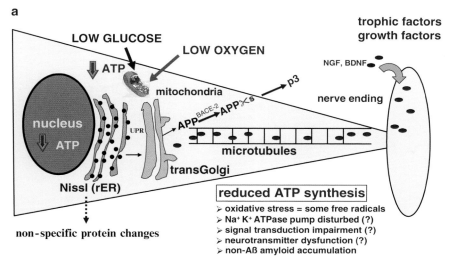

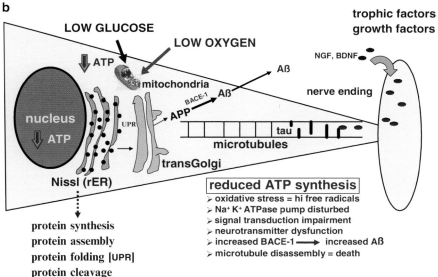

Chapter 13, Fig. 1 a A fundamental principle in cell biology is seen by the use of chemical energy in the form of ATP derived from glucose/oxygen delivery to assemble, disassemble, and alter protein structure. (For complete caption refer page 161)

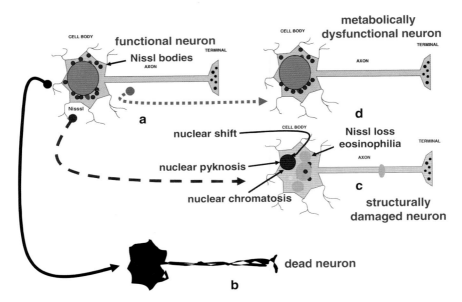

Chapter 13, Fig. 2 Functional and pathological states deranging brain neurons resulting in degenerative processes. Healthy, functional neurons (**a**) can either (**b**) die (*solid arrow*), (**c**) undergo metabolic-structural damage (*dashed arrow*), or (**d**) undergo only metabolic damage (*dotted arrow*). In the (**c**) state, intracellular damage can be assessed histologically by noting nuclear or cytoplasmic changes in organelles (chromatolysis, eosinophilia, nuclear shift, pyknosis, etc.). In the (**d**) state, biomolecular markers reflecting neuronal distress are needed to evaluate extent of metabolic dysfunction. Neuronal metabolic dysfunction following chronic brain hypoperfusion is likely the first step in the pathological pathway to cytostructural damage, atrophy, and death and how such dysfunction may explain cognitive disability, as in preclinical AD

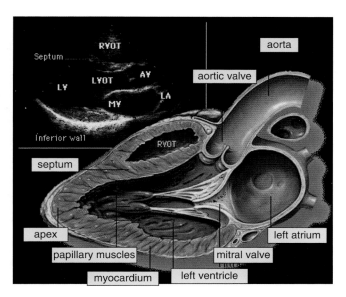

Chapter 13, Fig. 3 Echocardiography image (*top*) is obtained by placing the ultrasound probe on the surface of the chest over the heart. Image shows all four chambers of the heart, useful for detecting chamber enlargement. Valvular disease, hypertension, arrhythmias, left ventricular function, ejection fraction, and cardiac output can be determined and measured. These cardiac measurements offer a noninvasive, cost-effective, and reliable approach that can identify often correctable or treatable early lesions of the heart which can contribute to chronic brain hypoperfusion, cognitive impairment, and possibly AD in the elderly patient with mild memory complaints. Key: *RVOT* = right ventricular outlet tract; *LVOT* = left ventricular outlet tract; *LV* = left ventricle; *MV* = mitral valve; *LA*= left atrium

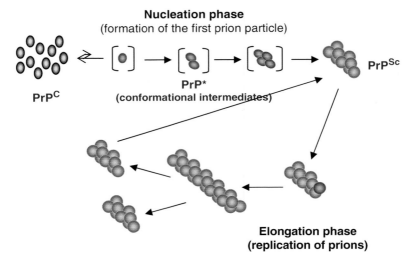

Chapter 14, Fig. 1 Seeding/nucleation hypothesis of prion protein misfolding and propagation. Prion replication as well as amyloid formation follow a process in which the limiting step is the stabilization of misfolded structures through formation of a minimal stable oligomer. Misfolded monomers are transient, getting cleared by the normal biological clearance pathways or returning to the natively folded conformation. When several misfolded monomers bind, they can form a stable oligomer that act as a seed to induce and stabilize further misfolded monomers, which are integrated into the growing polymer. The infectious agent acts as such in virtue of its capacity to serve as a nucleus to catalyze the process of protein misfolding and aggregation that result in the disease. Therefore, all protein misfolding processes following a seeding/nucleation mechanism have the inherent possibility to be infectious

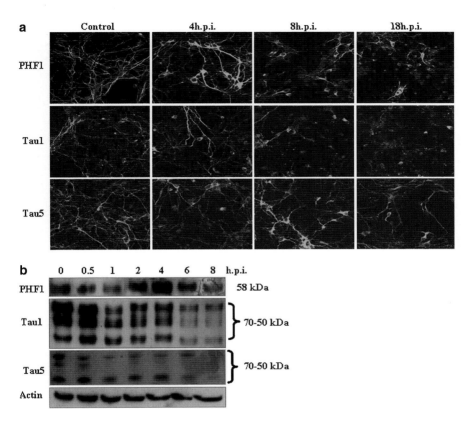

Chapter 15, Fig. 1 HSV-1 infection results in hyperphosphorylation of tau protein. **a** Primary cortical neurons infected with HSV-1 (moi 10) for 4, 8, and 18 h and untreated controls were stained with specific antibodies for PHF1, Tau1, and Tau5. Nuclei were counterstained with propidium iodide. The results are representative of three separate experiments. Magnification 100×. **b** Western blot analyses of PHF1, Tau1, Tau5, and actin in primary cortical neurons treated with HSV-1 (moi 10) at 0, 0.5, 1, 2, 4, 6, and 8 h post-infection (*hpi*). Blots shown are representative of three separate experiments. Reprinted from the *Journal of Alzheimer's Disease* (in press), Copyright (2008), with permission from IOS Press

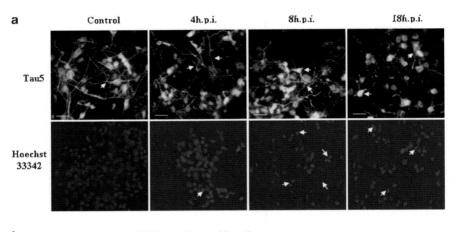

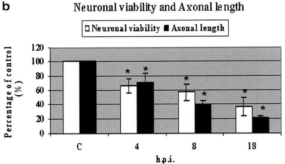

Chapter 15, Fig. 2 Neuronal cultures infected with HSV-1 exhibit axonal disruption. **a** Neurons infected with HSV-1 (moi 10) at 4, 8, and 18 h post-infection (*hpi*) and untreated controls were stained with a specific total tau antibody, Tau5, and the nuclei were counterstained with propidium iodide (*upper row*) or with Hoechst 33342 (*lower row*). **b** Neuronal viability and axonal length in HSV-1-infected cells were expressed at different times post-infection as a percentage of the values obtained in uninfected control. **c** Data represent mean ± SEM for three independent experiments (*$^{*}p < 0.05$ compared to untreated control). Reprinted from the *Journal of Alzheimer's Disease* (in press), Copyright (2008), with permission from IOS Press

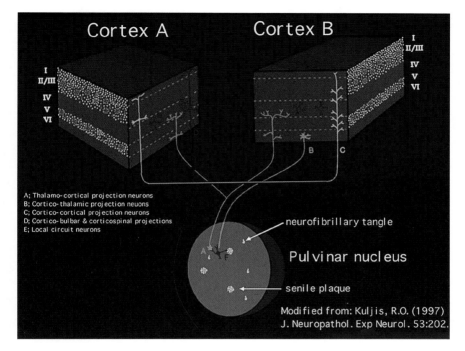

Figure content (labels within image):

Cortex A   Cortex B

A; Thalamo-cortical projection neurons
B; Cortico-thalamic projection neuons
C; Cortico-cortical projection neurons
D; Cortico-bulbar & corticospinal projections
E; Local circuit neurons

neurofibrillary tangle

Pulvinar nucleus

senile plaque

Modified from: Kuljis, R.O. (1997)
J. Neuropathol. Exp Neurol. 53:202.

Chapter 16, Fig. 1 Schematic rendering of the corticocortical and thalamocortical circuits affected by Alzheimer's disease. The *blocks* in the *upper part* of the figure represent parts of two different cortical areas. The *spheroid* in the *lower part* of the figure represents the pulvinar nucleus of the thalamus, one of the few in the entire thalamus that contain Alzheimer's disease lesions and is otherwise considered "unaffected." In all these representations, the *drop-like icons* signify neurofibrillary tangles, and the *semispheroidal yarn-like elements* signify senile plaques. *Roman numerals* represent cortical layers, and *regions with high densities of lesions* represent layers (V) or groups of layers (II/III) stereotypically affected in the disorder. With the exception of the pulvinar and a handful of very circumscribed regions, thalamocortical projections are essentially unaffected given the low density of lesions in layers IV and VI and in other subcortical regions. Similarly, the reciprocal corticothalamic projections (B) that originate primarily in layer VI also appear unaffected. In contrast, corticocortical connections that originate from medium-sized pyramidal neurons in layers II/III are distinctly and selectively affected given the high density of lesions in them. Other important elements in these circuits, which originate from layer V (labeled D in the figure), and some local circuit neurons (E, F) are also affected. Modified from Kuljis [13], with permission

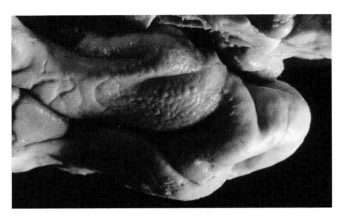

Chapter 16, Fig. 2 Modular organization of the entorhinal cortex. Photomacrograph of the medial aspect of the temporal lobe where the "glomeruli" or "warts" that decorate the surface of the entorhinal cortex are apparent and demarcate the extent of this area. From the author's collection

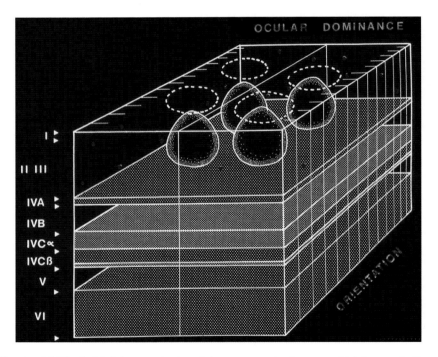

Chapter 16, Fig. 3 Modular organization of the macaque visual cortex, which is very similar to that in humans. Schematic representation of a segment of this area of the cortex as a cube in which the *upper surface* corresponds to the piamater and the *lower surface* corresponds to the interface between the gray and white matter. *Roman numerals* depict cortical layers and hatching regions with a high cytochrome oxidase (*CO*) activity. The *ovoid objects* in layers II/III correspond to the "patches" rich in CO that contain neurons that respond predominantly to wavelength (color) of visual stimuli as opposed to those responsive mainly to orientation in the interpatch spaces of the same layers. This wavelength versus orientation selectivity spans the thickness of the cortex in semicylindrical arrangements that form the basis of the "hypercolumns" postulated by Hubel and Wiesel [41] according to which this cortex consists of an array of hypercolumns. Modified from Kuljis and Rakic [39], with permission

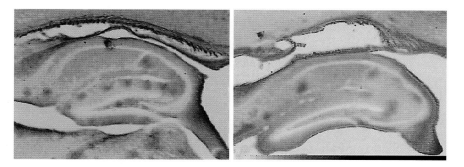

Chapter 16, Fig. 6 Photomicrographs of amyloid-containing-like lesions in the hippocampus of transgenic mice overexpressing normal human-α 1 antichymotrypsin. Gallyas' method for amyloid deposits (which is different from other methods from the same author to demonstrate myelin and cytoskeletal lesions). Note that the lesions in the hippocampus tend to have a selective laminar disposition, reflecting a propensity to affect certain elements of this structure, while sparing others in a manner similar to that in patients with Alzheimer's disease. Modified from Kuljis et al. [50]

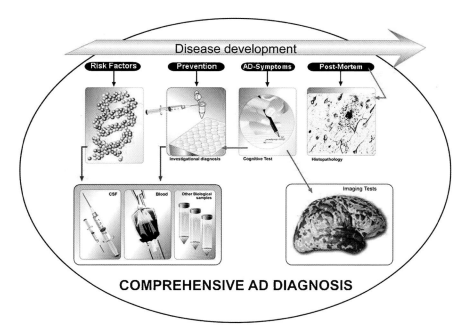

Chapter 18, Fig. 1 Current methods of Alzheimer's disease (*AD*) diagnosis such as cognitive tests and brain imaging are based on phenotypic changes, which are the consequence of irreversible brain deterioration. New approaches should aim to diagnose patients during the presymptomatic stages of the disease

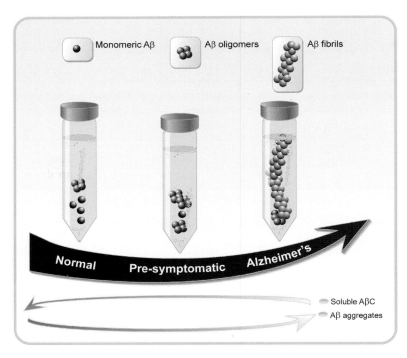

Chapter 18, Fig. 2 Amyloid-β (*Aβ*) misfolded oligomers may be circulating in biological fluids many years or even decades before the clinical symptoms of Alzheimer's disease appear. Moreover, their concentration might be indicative of the disease progression. One promising approach may be the specific detection of misfolded Aβ oligomers in biological fluids

Cerebral and Cardiac Vascular Pathology in Alzheimer's Disease

Jack C. de la Torre

Abstract A large body of evidence now indicates that Alzheimer's disease (AD) is a vascular disorder with neurodegenerative consequences and should be treated as such. Vascular risk factors in the elderly individual who already possesses a dwindling cerebrovascular reserve due to advancing age contribute to a further decline in cerebral blood flow (CBF) that results in unrelenting brain hypoperfusion. Brain hypoperfusion, in turn, gives rise to reduced ATP synthesis, thereby provoking a metabolic energy crisis involving ischemic-sensitive neurons and glial cells. Neuronal energy compromise accelerates oxidative stress, excess production of reactive oxygen species, aberrant protein synthesis, ionic pump dysfunction, signal transduction impairment, neurotransmitter failure, abnormal processing of amyloid-β protein precursor giving rise to amyloid-β deposition, and microtubule disruption from tau hyperphosphorylation. All high-energy metabolic changes leading to oxidative stress and cellular hypometabolism precede clinical expression of AD. Regional CBF measurements using neuroimaging techniques such as PET, SPECT, echocardiography, and Doppler ultrasound can help predict AD preclinically at the mild cognitive impairment stage or even before any clinical expression of the dementia is detected. Epidemiological studies together with findings from preclinical detection tools and present-day treatments for AD are proof of concept that AD is a vascular disorder that results in brain hypoperfusion. Both peripheral and cerebral vascular pathology can contribute to brain hypoperfusion. This new paradigm prompts redirection of our thinking and our efforts to decisively manage and treat 25 million people worldwide afflicted with this disorder.

J.C. de la Torre
10515 Santa Fe Drive
Sun City, AZ 85351
e-mail: jcdelatorre@comcast.net

R.B. Maccioni and G. Perry (eds.) *Current Hypotheses and Research Milestones in Alzheimer's Disease*. DOI: 10.1007/978-0-387-87995-6_13,
© Springer Science + Business Media, LLC 2009

1   Introduction

Immediate interruption of the cerebral energy supply is generally seen after cardiac arrest or stroke, two conditions that can result in global or local brain cell death. But what happens when brain energy supply dwindles over long periods of time following nonfatal cardiac disease or carotid-vertebral artery stenosis, two conditions that can lead to chronic cerebral hypoperfusion and are considered as major risk factors to Alzheimer's disease (AD)? When that happens, healthy, functional neurons can either die, undergo metabolic-structural damage, or undergo metabolic damage only, sparing cellular and molecular structural damage (Fig. 1).

The question is essential because glucose is the primary energy substrate for the brain which also consumes 20% of the total oxygen supply used by the body and its deficient delivery to the brain by a reduced blood flow can initiate a series of pathometabolic steps that can compromise brain cell function (Fig. 2a). We were led to ask this question in 1993, when the subject was only of rare interest, following our series of experimental brain hypoperfusion studies and a review of the clinical dementia literature at the time [1].

Since then, compelling evidence from a variety of clinical and basic studies has convinced us that AD should be treated as a vasocognopathy [2] (i.e., a vascular-related, cognitive disorder) and by so doing, the neurodegenerative subcellular products generated by reduced blood flow to the brain will be recognized as potential therapeutic targets that may respond reversibly to the energy crisis being generated (Fig. 2b). Extensive proof of concept now supports our vascular proposition. In this chapter, proof of concept is presented as a synopsis of selected findings from studies that provide evidence of a fundamental vascular connection in the pathophysiology of AD. These findings may offer a new directive for further research into this dementia.

Fig. 1 (continued) <fx1> = impaired reaction. **b** Persistent reduction of glucose and oxygen delivery to ischemic-sensitive neurons following chronic brain hypoperfusion leads to critical ATP depletion and promotes oxidative stress, reactive oxygen species, and abnormalities in protein synthesis activating their disassembly, misfolding (generating unfolded protein response, *UPR*), and abnormal cleavage. Altered protein synthesis can subsequently form molecules that can ravage nucleic acids, protein kinases, phosphatases, lipids, and harm cell structure. ATP cutback also hastens ionic pump dysfunction, signal transduction breakdown, neurotransmitter failure, faulty cleavage of amyloid-β protein precursor (*AβPP*) leading to BACE-1 upregulation and Aβ over-production, and microtubule damage from tau hyperphosphorylation. Retrograde axonal transport (curved arrow) of trophic and growth factors (e.g., NGF, BDNF) essential for neuronal survival is diminished due to energy-starved motor protein deficiency. This selective neuronal energy crisis is caused by blood flow supply not meeting cell energy demand in highly active brain regions whose vascular reserve capability has reached a critical threshold. About 75% of ATP energy is used on signaling (action potentials, postsynaptic potentials, etc.). Reduced ATP synthesis is the precursor of the molecular cascade that leads to AD and to the region-specific neuronal-glial death pathway. Key: ↓= reduction or loss; <fx1> = impaired reaction (*See Color Plates*)

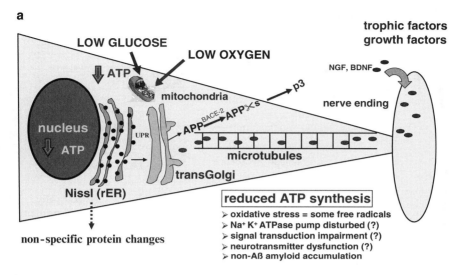

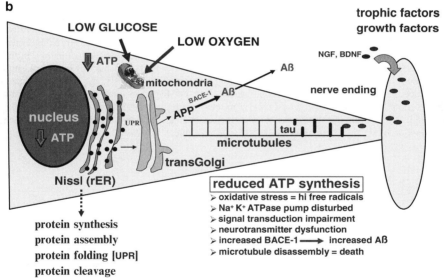

Fig. 1 a A fundamental principle in cell biology is seen by the use of chemical energy in the form of ATP derived from glucose/oxygen delivery to assemble, disassemble, and alter protein structure. Since proteins work to keep a neuron healthy, defects in their synthesis, assembly, folding, or cleavage can impair the normal intracellular and extracellular secretory transport pathway and damage brain cells. Normal neuronal energy metabolism in the brain includes optimal ATP production by mitochondrial oxidative phosphorylation, RNA transcription and protein synthesis, cell signaling, neurotransmission, and axonal transport of molecules between cytoplasm and nerve ending via microtubules. Normal brain has little or no amyloid-β (Aβ) accumulation. Following initial cerebral hypoperfusion, reduced ATP synthesis can result in nonspecific protein and oxidative stress changes that may lead to mild cognitive impairment. Theoretically, overexpression of Aβ peptides and specific neurodegenerative pathology may be lacking. Key: ↓= reduction or loss;

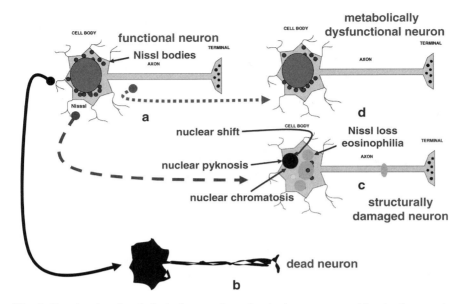

Fig. 2 Functional and pathological states deranging brain neurons resulting in degenerative processes. Healthy, functional neurons (**a**) can either (**b**) die (*solid arrow*), (**c**) undergo metabolic-structural damage (*dashed arrow*), or (**d**) undergo only metabolic damage (*dotted arrow*). In the (**c**) state, intracellular damage can be assessed histologically by noting nuclear or cytoplasmic changes in organelles (chromatolysis, eosinophilia, nuclear shift, pyknosis, etc.). In the (**d**) state, biomolecular markers reflecting neuronal distress are needed to evaluate extent of metabolic dysfunction. Neuronal metabolic dysfunction following chronic brain hypoperfusion is likely the first step in the pathological pathway to cytostructural damage, atrophy, and death and how such dysfunction may explain cognitive disability, as in preclinical AD (*See Color Plates*)

2 Epidemiology

Probably the most compelling evidence for a vascular connection to AD comes from independent epidemiological studies. Findings from the Rotterdam Study, the Kungsholmen project, EURODEM, FINMONICA, the PROCAM study, the Framingham study, and the Honolulu-Asia study, among others, indicate that over two dozen risk factors, all vascular-related, thus far have been recognized for AD (Table 1) [2–4]. These epidemiological studies have led to the conclusion that most of the AD cases analyzed have a vascular involvement, and that pure dementia types (including vascular dementia, VaD) in older subjects constitute only a minority of dementia cases [3, 4]. It is highly improbable that chance alone could explain the reduction of cerebral blood flow (CBF) induced by several dozen different risk factors found linked to AD. Building on the assumption that vascular risk factors for AD are valid and do lower CBF, we have introduced the role of *critically attained*

Table 1 Summary of findings from epidemiological studies (see text) of vascular-related risk factors to AD

Alzheimer's disease vascular risk factors
Brain-related risk factors
Aging
Ischemic stroke
Silent stroke
Head injury
Transient ischemic attacks
Migraine
Lower education
Hemorheological abnormalities
Depression
Circle of Willis atherosclerosis
Heart-related risk factors
Congestive heart failure
Valvular disease
Hypertension
Hypotension
Thrombotic episodes
High serum homocysteine
Atrial fibrillation
Presence of ApoE allele
Carotid atherosclerosis
Coronary artery bypass surgery
Peripheral risk factors
High serum cholesterol
High intake of saturated fat
Diabetes mellitus II
Hemorheological abnormalities
Alcoholism
Smoking
Menopause

Highly prevalent conditions such as stroke, hypertension, atherosclerosis, and heart disease are common precursors to AD

threshold of cerebral hypoperfusion (CATCH) [5]. CATCH develops when a vascular risk factor(s) further diminishes CBF in an individual with already declining cerebral perfusion due to advanced age and a poor vascular reserve. The reason is that CBF normally declines with age by about 21% from age 21 to 60 [6]. Below this normal age-related CBF decline, a "CATCH brain blood flow level" can result from advancing age when it coexists with a vascular risk factor as from the list in Table 1.

The speed at which CATCH is attained depends on the person's state of health, lifestyle, genetics, gender, diet, environmental exposure, and other confounding factors. Once CATCH is activated, energy hypometabolism and oxidative stress result from continuous brain hypoperfusion and from reduced glucose and oxygen delivery to neurons and glia [5] (Fig. 1b). The main reason for the high glucose/O_2

uptake in the normal brain is the massive amounts of ATP needed to maintain neuronal signaling during active ion channel flux. Ion flux is the basis for propagation of action potentials and neurotransmission. Impaired brain energy production also upregulates the amyloid-β (Aβ) rate-limiting enzyme BACE1, thus promoting Aβ overproduction [7] (Fig. 1a). Consequently, when mitochondrial function in neurons is quickly disrupted (as in stroke) by failing glucose/O_2 supply for energy production, rapid structural damage or death to brain cells occurs (Fig. 2). Hypometabolic compromise without apparent structural damage to neurons can be an alternative consequence when chronic brain hypoperfusion (as in atherosclerosis) is present for extended periods of time (Fig. 2). Brain hypoperfusion is the pathogenic trigger that pushes the neuronal energy crisis and the cascade of molecular and cytopathological changes that define AD (Fig. 1b).

Risk factors for AD are either the result of genetic susceptibility (e.g., carrying an apolipoprotein E ε4 allele) or environmental exposure of a person's health to an event that can introduce, accelerate, or further compromise cognitive dysfunction. At the present time, only environmental risk factor exposure is modifiable or preventable but in the future, genetic engineering of potential risk susceptibilities to AD could become a realistic therapeutic target for AD.

Vascular risk factors to AD as shown in Table 1 have been extensively reviewed [2–4, 8, 9] and only some recently reported risk factors will be briefly discussed here.

One example of a potentially modifiable vascular risk to AD of increasing research interest is hypercholesterolemia and dietary fat intake. A steady amount of research has confirmed a link between high levels of cholesterol and the development of AD [8].

Hypercholesterolemia acts as a precursor of atherosclerosis, cardiovascular disease, and diabetes. For obvious reasons, high serum cholesterol levels have generated a rapid-growing market for lipid-lowering drugs prescribed to patients at risk of cardiovascular or cerebrovascular conditions. Statins that lower cholesterol levels have been suggested as useful in both prevention and treatment of AD but this conclusion needs further verification [9]. The mechanism by which statins may provide a benefit against cerebrovascular disease including AD remains speculative and is likely multifactorial. In 1998, it was suggested, on the basis of murine studies, that the health benefits of statins did not merely aim at lowering lipids [10]. Statins were shown to protect against injury by a mechanism involving the selective upregulation of endothelial nitric oxide synthase [10], the enzyme that generates nitric oxide in blood vessels and is involved with regulating vascular function. Further research along these lines has demonstrated that statins improve endothelial function, increase nitric oxide bioavailability, provide antioxidant properties, inhibit inflammatory responses, exert immunomodulatory actions, regulate progenitor cells, and stabilize atherosclerotic plaques [11], and several studies have shown a benefit for statins in the treatment of AD [12].

Another important modifiable risk factor to AD is metabolic syndrome (MeS). MeS is a cluster of factors that can lead to the development of cardiovascular disease. MeS is characterized by abnormal insulin, glucose, and lipoprotein metabolism as well as by hypertension and obesity [13]. These factors are frequently found in people with either excess abdominal body fat or a decreased ability of the body to use insulin, which is known as insulin resistance [13]. MeS has been found to be associated with AD even when diabetic patients are excluded from analysis [14], suggesting that obesity associated with cardiovascular dysfunction or hyper-cholesterolemia is sufficient to initiate cognitive impairment that later can convert to AD. The number of people with MeS increases with age and is estimated to affect 40% of people beyond the age of 60. People with uncontrolled diabetes mel-litus type 2 and those with heart disease or prone to stroke are most likely to develop MeS [15].

Since obesity and lack of exercise are two conditions that can lead to MeS, correcting these by lifestyle changes that include a healthy diet and physical activity could help prevent or reverse MeS. The cardiovascular risk factors associated with MeS contribute to the development of atherosclerosis that can lead to a heart attack or a stroke. People with the MeS are also more likely to develop type 2 diabetes mellitus [15]. Insulin resistance is central to type 2 diabetes and is also implicated in the pathogenesis of AD [14]. This has prompted ongoing clinical trials in AD patients to test the efficacy of improving insulin-like signaling with dietary omega-3 fatty acids or insulin-sensitizing drugs as well as by exercise regimens.

It is well established that hypertension can increase the risk of stroke and heart problems and decrease life expectancy [2]. Many studies including the Framingham and the Honolulu-Asia Aging studies have implicated impaired cognitive function to hypertension in geriatric patients [16]. It has also been known for some time that hypertension in the elderly is a potential risk factor to AD [2]. What is not clear is how hypertension increases the incidence of AD, particularly in those not treated with antihypertensives. People with hypertension are six times more likely to have a stroke that can lead to AD [17]. It is calculated that reducing the number of cases affected by stroke which are amenable to prevention or treatment would have a major impact in lowering the incidence and societal costs of AD which is now estimated to affect 25 million people worldwide.

Some forms of cardiac disease that result in impairment of cardiac output or function with consequent reduction of cerebral perfusion have been found to increase AD risk [18]. Since 20% of cardiac output goes to the brain and 80% of carotid artery flow goes to the ipsilateral middle cerebral artery, it is no surprise that vascular lesions of the gray and white matter are frequently associated with cardiac disease and carotid artery occlusion, both vasculopathies that can induce AD.

Moreover, cerebral hypoperfusion can result in white matter lesions and cortical watershed microinfarcts in the absence of atherosclerosis or amyloid angiopathy [19], and may induce microinfarcts in the hippocampus prior to the onset of AD [20].

We firmly believe that AD onset can be delayed or possibly prevented if the heart can be kept healthy and brain hypoperfusion can be treated by restoring normal blood flow. The upshot would be a major research breakthrough in AD prognosis if either feat is successfully achieved. Consequently, diet, exercise, and management of many vascular risk factors amenable to treatment may be the key to significantly reducing the incidence of this dementia in the future.

3 Preclinical Detection of AD

The most important factor in developing a useful blueprint for AD treatment is identifying such candidates at the *preclinical* stage before any incapacitating cognitive damage has occurred. The reason is that once cognitive meltdown has begun (which implies serious neuronal damage and loss), AD pathology may be most difficult to reverse or arrest.

Considering AD as a vascular disorder characterized by brain hypoperfusion, techniques that can detect regional CBF reduction during mild cognitive impairment (MCI), a condition often seen prior to AD, or even before any memory deficits are measured, will be crucial to predict AD candidates.

In our judgment, three of the most important tools to screen AD candidates before major clinical symptoms have developed are single-photon emission tomography (SPECT), injected [^{18}F]fluorodeoxyglucose (FDG) using positron emission tomography (PET), and transcranial Doppler (TCD) ultrasonography. SPECT and TCD can detect with a high degree of sensitivity persons showing regional brain hypoperfusion at the MCI stage, many who later convert to AD [21, 22].

Preclinical detection of AD may now be possible using FDG-PET although the technique is considerably more expensive and labor intensive than SPECT or TCD. Individuals at risk of AD but who do not express any cognitive deficits show local hypometabolic reductions of the cerebral metabolic rate of glucose after FDG in brain regions that can later develop severe atrophy and neurodegeneration [23]. Since measures of glucose metabolic rate and CBF are almost linear in the resting state [24], FDG-PET studies are proof of concept that regional brain hypoperfusion precedes AD clinical symptoms and neurodegenerative changes. FDG-PET can also track AD progression [23].

Most patients with ischemic heart disease will present themselves in general practice. Therefore, the community management of ischemic heart disease has already become increasingly important and the role of the primary care physician even more pivotal.

Inasmuch as normalization of diminished cardiac output may prevent or reverse mild cognitive impairment [25], which is a form of chronic brain hypoperfusion that can lead to prodromal AD [5, 15, 18, 21, 28], simple, safe, and reliable diagnostic screens should be applied to older patients in an effort to reduce the consequences of cardiac to brain-related AD risk factors (Fig. 3). We have recommended the use of cost-effective, safe, and reliable noninvasive tools such as echocardiography and carotid Doppler ultrasound to be used in the elderly asymptomatic individual or those found to complain of memory difficulties [26, 27]. This clinical approach, whose acronym is "CAUSE" (carotid artery ultrasound-echocardiography), can detect cardiac pathology and extracranial vessel disease responsible for lowering CBF but are also potentially compliant to corrective medical or surgical treatment [27]. Since cardiac pathology and extracranial vessel disease have been shown to be major contributors to cerebral hypoperfusion and stroke, their attrition may result in a significant impact on lowering the prevalence of AD and VaD.

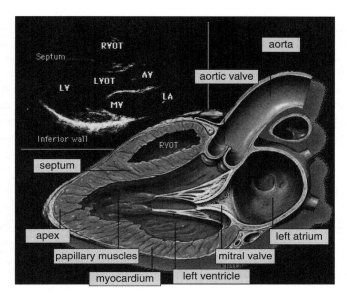

Fig. 3 Echocardiography image (*top*) is obtained by placing the ultrasound probe on the surface of the chest over the heart. Image shows all four chambers of the heart, useful for detecting chamber enlargement. Valvular disease, hypertension, arrhythmias, left ventricular function, ejection fraction, and cardiac output can be determined and measured. These cardiac measurements offer a noninvasive, cost-effective, and reliable approach that can identify often correctable or treatable early lesions of the heart which can contribute to chronic brain hypoperfusion, cognitive impairment, and possibly AD in the elderly patient with mild memory complaints. Key: *RVOT* = right ventricular outlet tract; *LVOT* = left ventricular outlet tract; *LV* = left ventricle; *MV* = mitral valve; *LA*= left atrium (*See Color Plates*)

4   Treatment of AD

Five drugs are available in the US market for prescriptive use in AD: Cognex (tacrine), Aricept (donepezil), Excelon (rivastigmine tartrate), and Reminyl (galantamine hydrobromide). All four act to slow the synaptic breakdown of acetylcholine, one of the neurotransmitters important in memory and learning. Reminyl also targets both AD and "mixed" dementia, that is, AD complicated by cerebrovascular pathology.

A fifth drug Namenda (memantine) is a newer medication for the treatment of AD that works ostensibly by antagonizing the N-methyl-D-aspartate receptors. Nonprescriptive treatments for AD in present use are nonsteroidal anti-inflammatory agents, ginkgo biloba, estrogen, vitamin E, and statins. All these treatments, whether prescriptive or over-the-counter, at best, provide only very modest symptomatic control generally at the early stages of AD and offer little to no benefit at the later stages of the disease. Their modest benefit to AD patients must be weighed against their frequent side effects, particularly the anticholinergic agents which include nausea, vomiting, dizziness, headaches, and depression.

It should be noted that the common link binding practically all the medicines and therapies available or tested experimentally for AD so far is their ability to *mildly* increase or improve CBF [28, 29]. However, since cerebral perfusion is only moderately improved, the beneficial effect of medicines such as those listed here is transient and modest. Interestingly, most of the products that improve AD symptoms also improve the symptoms of VaD. This activity by Pharmaca lend additional support to the concept of brain hypoperfusion in AD patients since increasing CBF in such patients, even mildly, does often result in modest symptomatic benefit.

5   Conclusions

AD is a vascular disorder with neurodegenerative consequences and should be treated as such. A large body of evidence including epidemiological, pharmaco-therapeutic, and neuroimaging studies support the concept of AD as a vascular disorder that results in brain hypoperfusion and a neuronal energy crisis that provokes the cellular and molecular changes that define this dementia. In this brief chapter, the many vascular-related risk factors to AD including some that are preventable are discussed. Considerable human data now point in a new direction for guiding future research into AD. This new research direction should open a window of opportunity for decisive management and treatment of this devastating disorder.

Acknowledgment   This research was supported by an Investigator-Initiated Research Grant from the Alzheimer's Association.

References

1. de la Torre JC, Mussivand T (1993) Can disturbed brain microcirculation cause Alzheimer's disease? Neurol Res 15:146–153
2. de la Torre JC (2004) Alzheimer's disease is a vasocognopathy: a new term to describe its nature. Neurol Res 26:517–524
3. Breteler MM (2000) Vascular risk factors for Alzheimer's disease: an epidemiologic perspective. Neurobiol Aging 21:153–160
4. Aguero-Torres H, Kivipelto M, von Strauss E (2006) Rethinking the dementia diagnoses in a population-based study: what is Alzheimer's disease and what is vascular dementia?. A study from the Kungsholmen project. Dement Geriatr Cogn Disord 22:244–249
5. de la Torre JC (2000) Critically attained threshold of cerebral hypoperfusion: the CATCH hypothesis of Alzheimer's pathogenesis. Neurobiol Aging 903:424–436
6. Leenders KL, Perani D, Lammertsma AA et al. (1990) Cerebral blood flow, blood volume and oxygen utilization. Normal values and effect of age. Brain 113:27–47
7. Velliquette RA, O'Connor T, Vassar R (2005) Energy inhibition elevates beta-secretase levels and activity and is potentially amyloidogenic in APP transgenic mice: possible early events in Alzheimer's disease pathogenesis. J Neurosci 25:10874–10883

8. Sjogren M, Mielke M, Gustafson D et al. (2006) Cholesterol and Alzheimer's disease – is there a relation? Mech Ageing Dev 127:138–147

9. Sparks DL, Sabbagh MN, Connor DJ (2005) Atorvastatin for the treatment of mild to moderate Alzheimer disease: preliminary results. Arch Neurol 62:753–757

10. Laufs U, La Fata V, Plutzky J, Liao JK (1998) Upregulation of endothelial nitric oxide synthase by HMG CoA reductase inhibitors. Circulation 97:1129–1135

11. Endres M, Laufs U, Huang Z (1998) Stroke protection by 3-hydroxy-3-methylglutaryl (HMG)-CoA reductase inhibitors mediated by endothelial nitric oxide synthase. Proc Natl Acad Sci USA 95:8880–8885

12. Wolozin B, Kellman W, Ruosseau P et al. (2000) Decreased prevalence of Alzheimer disease associated with 3-hydroxy-3-methyglutaryl coenzyme A reductase inhibitors. Arch Neurol 57:1439–1443

13. Grundy SM, Cleeman JI, Daniels SR, Donato KA (2005) Diagnosis and management of the metabolic syndrome: an American Heart Association/National Heart, Lung, and Blood Institute Scientific Statement. Circulation 112:2735–2752

14. Vanhanen M, Koivisto K, Moilanen L et al. (2006) Association of metabolic syndrome with Alzheimer disease: a population-based study. Neurology 67:843–847

15. Martins IJ, Hone E, Foster JK et al. (2006) Apolipoprotein E, cholesterol metabolism, diabetes, and the convergence of risk factors for Alzheimer's disease and cardiovascular disease. Mol Psychiatry 11:721–736

16. Elias MF, Wolf PA, D'Agostino RB et al. (1993) Untreated blood pressure level is inversely related to cognitive functioning: the Framingham Study. Am J Epidemiol 138:353–364

17. Ivan CS, Seshadri S, Beiser A (2004) Dementia after stroke: the Framingham Study. Stroke 35:1264–1268

18. Alves TC, Busatto GF (2006) Regional cerebral blood flow reductions, heart failure and Alzheimer's disease. Neurol Res 28:579–587

19. Miklossy J (2003) Cerebral hypoperfusion induces cortical watershed microinfarcts which may further aggravate cognitive decline in Alzheimer's disease. Neurol Res 25:605–610

20. de Leeuw FE, Barkhof F, Scheltens P (2004) White matter lesions and hippocampal atrophy in Alzheimer's disease. Neurology 62:310–312

21. Johnson KA, Jones K, Holman BL et al. (1998) Preclinical prediction of Alzheimer's disease using SPECT. Neurology 50:1563–1571

22. Ruitenberg A, den Heiker T, Bakker SL (2005) Cerebral hypoperfusion and clinical onset of dementia: the Rotterdam study. Ann Neurol 57:789–794

23. de Leon MJ, Convit A, Wolf OT (2001) Prediction of cognitive decline in normal elderly subjects with 2-[^{18}F]fluoro-2-deoxy-D-glucose/positron-emission tomography (FDG/PET). Proc Natl Acad Sci 98:10966–10971

24. Silverman DH, Phelps ME (2001) Application of positron emission tomography for evaluation of metabolism and blood flow in human brain: normal development, aging, dementia, and stroke. Mol Genet Metab 74:128–138

25. Vogels RL, Scheltens P, Schroeder-Tanka JM, Weinstein HC (2007) Cognitive impairment in heart failure: a systematic review of the literature. Eur J Heart Fail 9:440–449

26 de la Torre JC (2008) Alzheimer discase prevalence can be lowered with non-invasive testing. J. Alazheimers Dis 14:353–359

27 de la Torre JC (2008) Prevention of cognitive impairment and dementia with carotid artery ultrasound and echocardiography (CAUSE) Testing. J. Stroke and Cerebrovasc Dis

28. de la Torre JC (2002) Alzheimer's disease as a vascular disorder: nosological evidence. Stroke 33:1152–1162

29. de la Torre JC (2004) Is Alzheimer's a neurodegenerative or a vascular disorder? Data, dogma and dialectics. Lancet Neurol 3:184–190

Are Amyloids Infectious?

Rodrigo Morales, Baian Chen, and Claudio Soto

Abstract Misfolding and aggregation of proteins is the main feature of a group of maladies which include most of neurodegenerative diseases (such as Alzheimer's, Parkinson's, Huntington's, and prion diseases), as well as several systemic amyloidosis. Among them, prion diseases are the only one known to be infectious. Current evidence shows that the infectious agent in prion diseases is the misfolded protein and that the molecular mechanisms responsible for transmissibility are very similar to the process of amyloid formation in all protein misfolding disorders (PMDs). In this chapter, we discuss the theoretical and experimental evidences suggesting the possible infectious nature of several PMDs.

1 Introduction

DNA's linear code is translated into the three-dimensional information of proteins. This information is dictated by a proper folding process that depends on both the primary sequence of the polypeptide chain and the folding machinery present within the cell. Abnormal folding of proteins can lead to many pathological processes including loss of function or gain of toxic activity, which usually result from the accumulation of misfolded and aggregated proteins in specific tissues [1–3]. Protein misfolding disorders (PMDs) are a group of pathological conditions that include Alzheimer's disease (AD), Parkinson's disease (PD), amyotrophic lateral sclerosis, Huntington's disease, diabetes type-2, systemic amyloidosis, and prion diseases, among others [1–3]. A list of the diseases and proteins involved in PMDs is shown in Table 1.

R. Morales, B. Chen, and C. Soto (✉)
Protein Misfolding Disorders Laboratory, George and Cynthia Mitchell
Center for Neurodegenerative diseases
Department of Neurology, University of Texas Medical Branch
301 University Boulevard, Galveston, TX 77555
e-mail: clsoto@utmb.edu

R.B. Maccioni and G. Perry (eds.) *Current Hypotheses and Research Milestones in Alzheimer's Disease*. DOI: 10.1007/978-0-387-87995-6_14,
© Springer Science + Business Media, LLC 2009

Table 1 List of some protein misfolding disorders, the protein implicated, and the organ mostly affected by deposition of aggregates

Diseases	Protein involved	Affected organ
Alzheimer's disease	Amyloid-β protein, Tau	Brain
Type II diabetes	Islet amyloid polypeptide	Pancreas
Parkinson's disease	α-Synuclein	Brain
Primary amyloidosis (implicated in multiple myeloma, β-cell dyscracias)	Immunoglobulin light chain	Mostly kidney, liver, heart, and nerves
Huntington's disease	Huntingtin	Brain
Secondary or reactive amyloidosis	Amyloid-A	Mostly spleen, liver, and kidney
Spinocerebral ataxias	Ataxins	Brain
Transmissible spongiform encephalopathies	Prion protein	Brain
Hemodialysis-related amyloidosis	β2-Microglobulin	Bones and joints
Amyotrophic lateral sclerosis	Superoxide dismutase	Brain
Familial dementia of British or Danish type	ABri or ADan polypeptides	Brain
Senile systemic amyloidosis, familial amyloid polyneuropathy	Transthyretin	Heart, kidney, lungs, and peripheral nerves
Hereditary cerebral hemorrhage with amyloidosis Icelandic-type	Cystatin C	Brain
Familial amyloidosis, Finnish-type	Gelsolin	Peripheral and central nervous system
Familial amyloid polyneuropathy	Apolipoprotein A-I	Mostly in aorta
Senile amyloidosis	Apolipoprotein A-II	Multiple organs
Hereditary systemic amyloidosis, familial visceral amyloidosis	Lysozyme	Liver, spleen, and gastrointestinal tract
Serpin deficiency disorders (cirrhosis, angioedema)	Serpins	Liver and brain

It has been suggested that tissue deposition of misfolded protein aggregates might be responsible for cell impairment and death, leading subsequently to clinical symptoms in affected individuals. The insidious clinical symptoms, the progressive nature of the illness, and the lack of efficient therapeutic treatments for these diseases lead to severe problems for the quality of life of affected people and their families in both social and economic aspects. The expenses for the treatment and care of patients are very high and these costs progressively increase due to the severity of the disease. In addition, it is expected that the number of people that will be affected by these maladies will increase at a high rate during the coming years. Moreover, the lack of early diagnostic methods or effective treatments paints a bleak scenario for the future. For these reasons, it is urgent to move forward in trying to understand the mechanisms involved in the origin and development of these diseases. Exacerbating this state of affairs is a new hypothesis proposing that misfolded proteins could be infectious [4, 5]. In this chapter, we will discuss experimental evidence suggesting that PMDs could have an infectious origin, in a similar manner that is occurring for the transmissibility of prion diseases.

2 Common Features of Protein Misfolding Disorders

In spite of the important differences in clinical manifestation, PMDs share some common features such as their appearance late in life, the progressive and chronic nature of the disease, and the presence of deposits of misfolded protein aggregates [6]. These deposits are a typical disease signature and although in each disease the main protein component is different (Table 1), they have similar morphological, structural, and staining characteristics. Amyloid is the name originally given to extracellular protein deposits found in AD and systemic amyloid disorders [7], but it is nowadays used to refer in general to disease-associated protein aggregates [6]. All protein aggregates share similar structural and biochemical characteristics. Structurally, misfolding of proteins increases the level of β-sheet structure, leading to the formation of amyloid polymers organized as cross-β structures [8]. As a consequence, protein aggregates are resistant to proteolysis, denaturation, and general cellular clearance mechanisms.

The cellular factors and processes leading to the misfolded conformation have been partially identified. Among them, several mutations destabilizing the folded conformation and promoting its shift to the misfolded form have been identified in each protein [9]. These mutations usually result in dominant inheritance of the disease. Nevertheless, familial forms of the disease represent a small percentage of the total incidence of PMDs, and the majority of the cases have a sporadic origin. The discovery of mutations has led to the development of transgenic animal models overexpressing mutated proteins, where aggregates accumulation and cellular impairments occur in a similar form as in the human disease [10].

In spite of the key role of misfolded proteins in the disease, the mechanisms leading to cellular damage and tissue dysfunction are still unclear. The toxicity of protein aggregates has been extensively documented in vitro and in vivo [3, 11, 12] However, it is still not clear which type of aggregates are the most toxic species. Recent evidence supports the hypothesis that smaller and soluble oligomeric aggregates on pathway to form the large fibrillar deposits could be the molecules mostly responsible for the toxic effects observed in these maladies [12–14]. Fibers could be acting as a protective mechanism in order to trap these particles and encapsulate them in the tissue.

3 Prion Diseases: The Bizarre Infectious Member of the PMD Group

Prion diseases, or transmissible spongiform encephalopathies (TSEs), are a group of rare, fatal neurodegenerative diseases, affecting humans and several species of mammals [15]. The most prevalent form of human TSE is Creutzfeldt-Jakob disease (CJD) and the most common animal prion disease is scrapie affecting sheep. However, the most worrisome TSEs are the new diseases, variant CJD in humans, bovine spongiform encephalopathy in cattle, and chronic wasting disease in cervids. According to how they arise, prion diseases can be classified as sporadic, familial, or infectious. Transmission of prion diseases was first reported in 1937 when sheep were inoculated with a vaccine prepared from formalin-treated sheep brain [16]. Since then, the search for the TSE infectious agent was (and still is) a matter of great effort from many investigators worldwide [17]. At this time, experimental evidence points out that the only component of the infectious agent in TSEs is the misfolded prion protein (referred as PrP^{Sc}) [17–19]. Highly purified preparations of PrP^{Sc} are infectious and the quantity of this protein correlates very well with infectivity [20]. Supporting this information, transgenic animals lacking the normal version of the prion protein (PrP^{C}) are completely resistant to infection, whereas animals overexpressing this protein are more susceptible [21, 22].

Prion diseases have been identified and transmitted to several animal species. The development of transgenic mice expressing PrP from different TSE sensitive mammals has helped to understand prion propagation processes among relevant species such as sheep, cattle, cervids, pigs, and humans [23]. Using these models, several features of the pathology have been studied. One interesting example is related with the transmission of disease to animal models by inoculation with different tissues and body fluids. The presence of the infectious material in body fluids presents an alarming scenario.

The central concept in the prion hypothesis is that PrP^{Sc} is the only component of the infectious agent, which can "replicate" in the brain in the absence of nucleic acids by converting the natively folded PrP^{C} into the misfolded form [18, 19]. Prion replication is hypothesized to occur when PrP^{Sc} in the infecting inoculum interacts specifically with host PrP^{C}, catalyzing its conversion to the pathogenic form of the protein. The precise molecular mechanism of PrP^{C} → PrP^{Sc} conversion is not well understood. However, the available data support the seeding/nucleation model in which infectious PrP^{Sc} is an oligomer that acts as a seed to bind PrP^{C} and catalyze its conversion into the misfolded form by incorporation into the growing polymer (Fig. 1) [24, 25]. In this model, in the absence of seeds, the spontaneous PrP conversion process will be thermodynamically unfavorable and kinetically very slow, which could account for the extremely low incidence of sporadic TSEs. However, the addition of enough seeds by infection leads to disease in 100% of the cases.

The recent generation of infectious prions in vitro has been an excellent support for the protein-only hypothesis of prion misfolding and replication. Recently, Legname et al. showed the generation of infectious prions in the test tube. In this study, they used a specially folded fragment (89–230) of the recombinant mouse prion protein [26]. The final folded product results in similar biochemical properties to PrP^{Sc} as assessed by resistance to proteases and insolubility. This preparation was inoculated in transgenic animals overexpressing PrP and the animals developed a spongiform encephalopathy characterized by long incubation periods and the presence of proteinase-K-resistant PrP. Although this report provides interesting data, several problems with their experimental design limit its scope as the final proof for the prion hypothesis. Using a different strategy, our group has developed an in vitro system to replicate prions. This technique, termed Protein Misfolding Cyclic Amplification, is able to replicate in vitro the misfolded form of PrP^{Sc} at

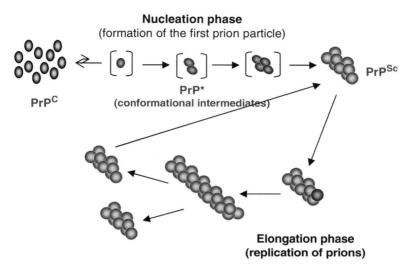

Fig. 1 Seeding/nucleation hypothesis of prion protein misfolding and propagation. Prion replication as well as amyloid formation follow a process in which the limiting step is the stabilization of misfolded structures through formation of a minimal stable oligomer. Misfolded monomers are transient, getting cleared by the normal biological clearance pathways or returning to the natively folded conformation. When several misfolded monomers bind, they can form a stable oligomer that act as a seed to induce and stabilize further misfolded monomers, which are integrated into the growing polymer. The infectious agent acts as such in virtue of its capacity to serve as a nucleus to catalyze the process of protein misfolding and aggregation that result in the disease. Therefore, all protein misfolding processes following a seeding/nucleation mechanism have the inherent possibility to be infectious (*See Color Plates*)

expenses of the normally produced PrP^C [27, 28]. This assay, based on the seeding/ nucleation model, is able to generate large amounts of infectious material using a small amount of starting PrP^{Sc}. Experimentally, this process is performed by mixing minute amounts of infectious prions in brain homogenate from healthy animals. Using this technique we reported the successful amplification of hamster prions [27, 28]. By serial dilutions of the in vitro generated material in healthy brain homogenates, we were able to dilute out the original inoculum obtaining only in vitro generated prions. Biochemical and structural analysis of the newly generated misfolded proteins showed similar characteristics to the original material. Importantly, in vivo challenge of this preparation to wild-type hamsters showed that these prions were infectious producing a similar disease compared to the original prions [28].

4 Are Other Protein Misfolding Disorders Infectious?

The misfolding and aggregation pathway, its mechanism, and structural intermediates are very similar in all PMDs, including TSEs [4, 5]. The mechanism by which amyloid is formed is similar to the process of prion replication. The seeding/nucleation model accounts for the generation of infectious PrPSc and for the formation of amyloid fibrils. In vitro experiments have shown that most PMD-associated proteins form aggregates with kinetic features of nucleated polymerization [29, 30]. The seeding-nucleation model provides a rationale and plausible explanation for the infectious nature of prions and suggests that protein misfolding processes as those associated with several human diseases have the inherent ability to be transmissible [4, 5]. Infectivity lies on the capacity of preformed stable misfolded oligomeric proteins to act as a seed to catalyze the misfolding and aggregation process (Fig. 1) [4, 5, 31]. The acceleration of protein aggregation by addition of seeds has been convincingly reported in vitro for several proteins implicated in diverse PMDs [6, 32, 33]. Extrapolating the in vitro results to the in vivo situation, the correct administration of preaggregated, stable misfolded structures should substantially accelerate the misfolding, aggregation, and tissue accumulation of the protein. Provided that protein misfolding and aggregation is the cause of the disease, this should lead to the acceleration of a pathogenic process that in the absence of the seed was set to occur much later in life or not at all during the life span of the individual.

Considering the knowledge gained in TSEs, it is likely that infection by misfolded proteins will produce a very long incubation period between the times of infection and the manifestation of clinical disease. For this reason, a possible infectious origin would be very hard to identify. The recognition of TSEs as infectious was possible principally due to a couple of fortuitous events: the use of contaminated material to treat a specific flock of sheep in the 1930s [16] and the discovery of Kuru transmission by cannibalistic rituals [34]. The rare prevalence of TSEs makes it easy to identify an isolated infectious event. However, in more frequent diseases, such as AD, PD, or diabetes, it would be difficult to associate disease with a possible infectious exposure that occurred decades before. Nevertheless, the mechanistic similarities between TSEs and other PMDs have prompted investigators to search for this possibility [4, 5, 35]. As a result of these studies, there are several pieces of evidence in favor and against the transmissibility of misfolding related disorders. Next, we will discuss some of these studies in three PMDs where the literature is most abundant.

4.1 Alzheimer's Disease

The idea that AD can be infectious has been around for a long time. The prime approaches to prove this hypothesis consisted in the transmission of TSE by injecting brain homogenates from AD patients to nonhuman primates; however, the results of these experiments were conflicting, with some of them positive [36] and some negative [37]. One possible reason for this is the low number of primates used for these studies. In addition, a possible "species-barrier" effect, similar to what is observed in prions [38], could also be responsible for the conflicting results. Altogether, the limited number of animals tested and the lack of knowledge about prion-related transmission at the time the experiments were performed suggest that these studies should be repeated using optimal experimental conditions.

With the emergence of transgenic mice models expressing the human amyloid protein and developing some of the neuropathological, biochemical, and clinical characteristics of AD [39], the search for a possible infectious origin for the disease took a new breath. Recently, Kane et al. showed that the inoculation of brain homogenates from AD patients to the Tg2576 transgenic model can accelerate the appearance of amyloid plaques [40, 41]. In these studies, unilateral inoculation of AD preparations accelerated the aggregation process 5 months postinoculation, but only in the inoculated hemisphere [40]. After 12 months, experimental animals presented amyloid-β (Aβ) accumulation in both hemispheres but the Aβ load was clearly greater in the injected one [41]. A followed-up study reported that the seeding activity of brain extracts was reduced or abolished by Aβ immunodepletion, protein denaturation, or Aβ immunization [42]. Interestingly, the phenotype of the exogenously induced amyloidosis depended on both the characteristics of the host and the source of the agent. These findings indicate that AD brain preparations can accelerate the aggregation process. However, since spontaneous accumulation of Aβ occurs in these animals at a later time, it is not possible to state that AD misfolded aggregates are infectious, but just accelerating a process that was set to occur because of the genetic manipulation of the mice.

4.2 Reactive Systemic Amyloidosis

Reactive systemic amyloidosis is a potential complication of any disorder that gives rise to a sustained acute-phase inflammatory response [43]. The list of chronic inflammatory, infective, or neoplastic disorders that can underlie it is almost without limit. The prevalence of amyloid deposition in patients with chronic inflammatory diseases is 3.6–5.8%, being the most common rheumatoid arthritis [44]. The protein deposited in this disease is called amyloid-A (AA) and is derived from cleavage fragments of the circulating acute-phase reactant serum AA protein (SAA) [45]. SAA is an apolipoprotein of high-density lipoprotein which is synthesized by hepatocytes. The molecular weight is 11.4–12.5 kDa in different species and its expression is regulated by cytokines. SAA is usually at a low concentration in the plasma (~20 mg/liter) but its concentration can increase up to a 1000-fold under inflammatory conditions [46]. Although AA amyloidosis can develop rapidly, the median latency between presentation with a chronic inflammatory disorder and clinically significant amyloidosis is almost two decades. AA aggregates can be found in tissues such as liver, spleen, and kidneys [47]. It has been found that experimental induction of inflammation followed by challenge with tissue homogenates containing AA seeds can produce a clear shortening in the lag phase of protein accumulation compared with non-AA seed-treated animals [48]. It has been established that the active principles of these preparations, referred to as amyloid enhancing factor (AEF), are indeed the misfolded units of AA [49]. As in prion diseases, the oral challenge of AEF in minute amounts results in a decrease in the AA deposition time suggesting AEF is very resistant to degradation and elimination. Interestingly,

AEF-propagating properties are also completely abolished after treatment with dena-turing agents [50]. While AEF shares many properties with prions, its properties, however, as an "enhancer" and not transmissible factor per se make it difficult to consider AA aggregates as a bona fide infectious agent.

4.3 Mouse Senile Amyloidosis

The accumulation of apolipoprotein AII (apoAII) in some strains of old mice leads to senile amyloidosis [51]. Young animals do not accumulate apoAII fibrils, but progressively with age, animals begin to accumulate misfolded apoAII aggregates in diverse organs during aging [52]. It has been reported that inoculation of pre-formed apoAII seeds leads to the aggregation of this protein in young mice [53]. In contrast, no effects were observed after the administration of denatured apoAII preparations [54]. The "infective" effects of these aggregates were also reported after oral challenge [55]. Surprisingly, the fact that apoAII aggregates are present in the feces of the affected animals suggests a possible mechanism for the transmis-sion of the disease by this route [55]. In order to answer this question, untreated animals were housed in the same cage with apoAII-affected mice. It was found that untreated mice generated amyloidosis at early stages, just by sharing the cage with old animals. The oral transmission via feces was suggested because p.o. administration of feces from old animals into young animals induced disease [55]. Moreover, transmission of apoAII amyloidosis exhibits a "strain phenomenon" analogous to the prion strains [56]. This data suggests that apoAII misfolded aggregates might be really a prion-type of infectious agent.

Acknowledgments We thank Dr. Kristi Green for critical review of the manuscript. This manu-script was supported by a grant from the CART Foundation to Claudio Soto.

References

1. Soto C (2001) Protein misfolding and disease; protein refolding and therapy. FEBS Lett 498:204–207
2. Chiti F and Dobson CM (2006) Protein misfolding, functional amyloid, and human disease. Annu Rev Biochem 75:333–366
3. Carrell RW (2005) Cell toxicity and conformational disease. Trends Cell Biol 15:574–580
4. Soto C, Estrada L, Castilla J (2006) Amyloids, prions and the inherent infectious nature of misfolded protein aggregates. Trends Biochem Sci 31:150–155
5. Walker LC, Levine H III, Mattson MP, Jucker M (2006) Inducible proteopathies. Trends Neurosci 29:438–443
6. Soto C (2003) Unfolding the role of protein misfolding in neurodegenerative diseases. Nature Rev Neurosci 4:49–60
7. Glenner GG (1980) Amyloid deposits and amyloidosis. The beta-fibrilloses (first of two parts). N Engl J Med 302:1283–1292

8. Blake CC, Serpell LC, Sunde M, Sandgren O, Lundgren E (1996) A molecular model of the amyloid fibril. Ciba Found Symp 199:6–15

9. Hardy J, Gwinn-Hardy K (1998) Genetic classification of primary neurodegenerative disease. Science 282:1075–1079

10. Rockenstein E, Crews L, Masliah E (2007) Transgenic animal models of neurodegenerative diseases and their application to treatment development. Adv Drug Deliv Rev 59:1093–1102

11. Canevari L, Abramov AY, Duchen MR (2004) Toxicity of amyloid beta peptide: tales of calcium, mitochondria, and oxidative stress. Neurochem Res 29:637–650

12. Haass C, Selkoe DJ (2007) Soluble protein oligomers in neurodegeneration: lessons from the Alzheimer's amyloid beta-peptide. Nat Rev Mol Cell Biol 8:101–112

13. Glabe CG (2006) Common mechanisms of amyloid oligomer pathogenesis in degenerative disease. Neurobiol Aging 27:570–557

14. Caughey B, Lansbury PT (2003) Protofibrils, pores, fibrils, and neurodegeneration: separating the responsible protein aggregates from the innocent bystanders. Annu Rev Neurosci 26:267–298

15. Collinge J (2001) Prion diseases of humans and animals: their causes and molecular basis. Annu Rev Neurosci 24:519–550

16. Cullie J, Chelle PL (1939) Experimental transmission of trembling to the goat. Comptes Rendus des Seances de l'Academie des Sciences 208:1058–1160

17. Aguzzi A, Polymenidou M (2004) Mammalian prion biology: one century of evolving concepts. Cell 116:313–327

18. Soto C, Castilla J (2004) The controversial protein-only hypothesis of prion propagation. Nat Med 10:S63–S67

19. Prusiner SB (1998) Prions. Proc Natl Acad Sci USA 95:13363–13383

20. Prusiner SB (1982) Novel proteinaceous infectious particles cause scrapie. Science 216:136–144

21. Bueler H, Aguzzi A, Sailer A, et al. (1993) Mice devoid of PrP are resistant to scrapie. Cell 73:1339–1347

22. Fischer M, Rülicke T, Raeber A, et al. (1996) Prion protein (PrP) with amino-proximal deletions restoring susceptibility of PrP knockout mice to scrapie. EMBO J 15:1255–1264

23. Groschup MH, Buschmann A (2008) Rodent models for prion diseases. Vet Res 39:32.

24. Caughey B (2003) Prion protein conversions: insight into mechanisms, TSE transmission barriers and strains. Br Med Bull 66:109–120

25. Soto C, Saborio GP, Anderes L (2002) Cyclic amplification of protein misfolding: application to prion-related disorders and beyond. Trends Neurosci 25:390–394

26. Legname G, Baskakov IV, Nguyen HO, et al. (2004) Synthetic mammalian prions. Science 305:673–676

27. Saborio GP, Permanne B, Soto C (2001) Sensitive detection of pathological prion protein by cyclic amplification of protein misfolding. Nature 411:810–813

28. Castilla J, Saá P, Hetz C, Soto C (2005) In vitro generation of infectious scrapie prions. Cell 121:195–206

29. Jarrett JT, Lansbury PT Jr (1993) Seeding "one-dimensional crystallization" of amyloid: a pathogenic mechanism in Alzheimer's disease and scrapie?. Cell 73:1055–1058

30. Walker LC, Bian F, Callahan MJ, et al. (2002) Modeling Alzheimer's disease and other proteopathies in vivo: is seeding the key?. Amino Acids 23:87–93

31. Gajdusek DC (1994) Nucleation of amyloidogenesis in infectious and noninfectious amyloidoses of brain. Ann NY Acad Sci 724:173–190

32. Harper JD, Lansbury PT Jr (1997) Models of amyloid seeding in Alzheimer's disease and scrapie: mechanistic truths and physiological consequences of the time-dependent solubility of amyloid proteins. Annu Rev Biochem 66:385–407

33. Krebs MR, Morozova-Roche LA, Daniel K, Robinson CV, Dobson CM (2004) Observation of sequence specificity in the seeding of protein amyloid fibrils. Protein Sci 13:1933–1938

34. Gajdusek DC, Gibbs CJ, Alpers M (1966) Experimental transmission of a Kuru-like syndrome to chimpanzees. Nature 209:794–796

35. Sigurdsson EM, Wisniewski T, Frangione B (2002) Infectivity of amyloid diseases. Trends Mol Med 8:411–413

36. Baker HF, Ridley RM, Duchen LW, Crow TJ, Bruton CJ (1994) Induction of beta (A4)-amyloid in primates by injection of Alzheimer's disease brain homogenate. Comparison with transmission of spongiform encephalopathy. Mol Neurobiol 8:25–39
37. Goudsmit J, Morrow CH, Asher DM, et al. (1980) Evidence for and against the transmissibility of Alzheimer disease. Neurology 30:945–950
38. Morales R, Abid K, Soto C (2007) The prion strain phenomenon: molecular basis and unprecedented features. Biochim Biophys Acta 1772:681–691
39. Duff K (2001) Transgenic mouse models of Alzheimer's disease: phenotype and mechanisms of pathogenesis. Biochem Soc Symp 67:195–202.
40. Kane MD, Lipinski WJ, Callahan MJ, et al. (2000) Evidence for seeding of beta-amyloid by intracerebral infusion of Alzheimer brain extracts in beta-amyloid precursor protein-transgenic mice. J Neurosci 20:3606–3611
41. Walker LC, Callahan MJ, Bian F, et al. (2002) Exogenous induction of cerebral beta-amyloidosis in betaAPP-transgenic mice. Peptides 23:1241–1247
42. Meyer-Luehmann M, Coomaraswamy J, Bolmont T, et al. (2006) Exogenous induction of cerebral beta-amyloidogenesis is governed by agent and host. Science 313:1781–1784
43. Sipe JD (2000) Serum amyloid A: from fibril to function. Current status. Amyloid 7:10–12
44. Buxbaum JN (2004) The systemic amyloidoses. Curr Opin Rheumatol 16:67–75
45. Anders RF, Natvig JB, Michaelsen TE, Husby G (1975) Isolation and characterization of amyloid-related serum protein SAA as a low molecular weight protein. Scand J Immunol 4:397–401
46. De Beer FC, Mallya RK, Fagan EA, Lanham JG, Hughes GR, Pepys MB (1982) Serum amyloid-A protein concentration in inflammatory diseases and its relationship to the incidence of reactive systemic amyloidosis. Lancet 2:231–234
47. Sipe JD, Colten HR, Goldberger G, et al. (1985) Human serum amyloid A (SAA): biosynthesis and postsynthetic processing of preSAA and structural variants defined by complementary DNA. Biochemistry 24:2931–2936
48. Kisilevsky R, Boudreau L (1983) Kinetics of amyloid deposition. I. The effects of amyloid-enhancing factor and splenectomy. Lab Invest 48:53–59
49. Ganowiak K, Hultman P, Engstrom U, Gustavsson A, Westermark P (1994) Fibrils from synthetic amyloid-related peptides enhance development of experimental AA-amyloidosis in mice. Biochem Biophys Res Commun 199:306–312
50. Lundmark K, Westermark GT, Nyström S, Murphy CL, Solomon A, Westermark P (2002) Transmissibility of systemic amyloidosis by a prion-like mechanism. Proc Natl Acad Sci USA 99:6979–6984
51. Higuchi K, Naiki H, Kitagawa K, et al. (1995) Apolipoprotein A-II gene and development of amyloidosis and senescence in a congenic strain of mice carrying amyloidogenic ApoA-II. Lab Invest 72:75–82
52. Higuchi K, Naiki H, Kitagawa K, Hosokawa M, Takeda T (1991) Mouse senile amyloidosis. ASSAM amyloidosis in mice presents universally as a systemic age-associated amyloidosis. Virchows Arch B Cell Pathol Incl Mol Pathol 60:231–238
53. Xing Y, Higuchi K (2002) Amyloid fibril proteins. Mech Ageing Dev 123:1625–1636
54. Higuchi K, Kogishi K, Wang J, et al. (1998) Fibrilization in mouse senile amyloidosis is fibril conformation-dependent. Lab Invest 78:1535–1542
55. Xing Y, Nakamura A, Chiba T, et al. (2001) Transmission of mouse senile amyloidosis. Lab Invest 81:493–499
56. Xing Y, Nakamura A, Korenaga T, et al. (2002) Induction of protein conformational change in mouse senile amyloidosis. J Biol Chem 277:33164–33169

The Possible Link Between Herpes Simplex Virus Type 1 Infection and Neurodegeneration

Carola Otth, Angara Zambrano, and Margarita Concha

Abstract Herpes simplex virus type 1 and type 2 (HSV-1 and HSV-2) are ubiquitous, neurotropic, and the most common pathogenic cause of sporadic acute encephalitis in humans. Herpes simplex encephalitis is associated with a high mortality rate and significant neurological sequelae, which afflict patients for life. HSV-1 has been suggested as an environmental risk factor for Alzheimer's disease. However, the link between HSV-1 infection and neurodegenerative processes is still unclear. It has been proposed that the innate immune response to the virus particularly the activation of Toll-like receptor pathways in astrocytes and microglia could lead to neurodegeneration. Finally, we have also shown that in vitro HSV-1 neuronal infection triggers a change in the hyperphosphorylation state of tau and also results in marked neuritic damage and neuronal death.

1 Herpes Simplex Virus Type 1

Herpes simplex virus type 1 (HSV-1) infection usually occurs during childhood, typically as a result of the direct inoculation of infected droplets from orolabial or nasal secretions onto susceptible mucosal surfaces. Subsequently, during initial productive infection with HSV-1, the virus reaches the cell bodies of the sensory and sympathetic neurons in the trigeminal ganglia through retrograde transport, establishing latent infection. Diverse non-specific events, including inflammation, can trigger HSV-1 reactivation. Reactivated virus is transported back to the body surface to cause recurrent lesions manifested as cold sores (*herpes labialis*). HSV-1 infection is virtually universal – most adults are seropositive – however, only 20–40% of infected people develop symptoms [1]. Moreover, HSV-1 remains lifelong in the peripheral neurons

C. Otth (✉), A. Zambrano, and M. Concha
Instituto de Microbiología Clínica, Facultad de Medicina, Universidad Austral de Chile
Casilla, P.O. Box 567, Valdivia, Chile
e-mail: cotth@uach.cl

R.B. Maccioni and G. Perry (eds.) *Current Hypotheses and Research Milestones in Alzheimer's Disease*. DOI: 10.1007/978-0-387-87995-6_15,
© Springer Science + Business Media, LLC 2009

181

of the infected host in a latent state, occasionally it can also be transported to the central nervous system (CNS) causing rapidly progressive, necrotizing, and fatal encephalitis in humans [2]. Even with the advent of anti-viral therapy, effective treatments for HSV-1 brain infection are limited because the cause of the resulting neuropathogenesis is not completely understood. Herpes simplex encephalitis (HSE) has an estimated annual incidence in the population of 1 in 250,000 to 1 in a million [2] and causes high mortality as well as significant morbidity in survivors. Even though treatment of HSE with acyclovir in the early stages of the disease decreases mortality, patients are not free of further complications [2]. In fact, many suffer from severe cognitive impairments which can seriously affect their quality of life. In younger people, the virus is almost always absent in brain [3]; therefore, it has been suggested that HSV-1 reaches the brain in older age, when the immune system declines, and resides there in latent form until stress or immunosuppression reactivates it, causing an acute but localized infection – a 'mild' encephalitis [4].

2   HSV-1 and Neurodegenerative Process

Several studies have suggested that HSV-1 should be considered as an environmental risk factor for Alzheimer's disease (AD) [4–7], in part because in HSE, the main regions affected are the temporal and frontal cortices and the hippocampus [5]; therefore, these patients have increased risk of memory loss and present deterioration of intellectual and mental capabilities, and also because HSV-1 DNA can be detected in the brain of normal aged individuals as well as in AD patients [7–9]. Moreover, it was shown that the presence of HSV-1 in the brain of patients carrying the type 4 allele of the apolipoprotein E gene (APOE), which is a well-known susceptibility factor for AD, confers a higher risk to develop this disease [7, 10]. Therefore, studies that help to elucidate the role of HSV-1 in AD are urgently needed.

It is plausible that reactivation of latent HSV-1 in the CNS can cause local and regional neurodegenerative injury. Previously, HSV-1 DNA and viral antigens were found in brain and neurons of AD patients, suggesting HSV-1 reactivation in these neurons [11]. Although the pathogenesis of HSV encephalitis in humans is not well known, the possibility of reactivation of latent virus in brain tissue has not been excluded [2]. However, direct evidence conclusively showing that HSV-1 reactivates in neurons and causes local neurodegenerative damage is still lacking. Hill et al. [10] suggested the possibility that age-associated neurodegeneration could be caused by inflammation and cell death, which could have started with reactivation of HSV. Microtubular dynamics is important in all neuronal functions, such as vesicular traffic and synaptic connections, and it can be affected during neurodegenerative processes or as a result of viral infections. Tau is a microtubule-associated protein expressed mainly in neurons; it is encoded by a single gene, but because of alternative splicing and phosphorylation, it shows multiple isoforms. These isoforms differ in the number of conserved repeats (three or four), located in tandem in the C-terminal region and in the number of insertions (none, one, or two) in the N-terminal portion of the protein [12].

These repeats constitute tau-binding domains to β-tubulin and by this means promote microtubule assembly, but it may also participate in other cellular processes such as linking signal-transduction pathways to the cytoskeleton. Tau is functionally modulated by phosphorylation and is highly phosphorylated in several neurodegenerative diseases [13, 14]. In addition, it has been demonstrated that the ability of tau to bind to and to stabilize microtubules correlates inversely with its phosphorylation state, which may in turn facilitate its self-assembly. Neurofibrillary tangles constitute the major hallmark present in pathological processes associated to neurodegenerative diseases such as AD, tauopathies, and Parkinson's disease. These intracellular aggregates are composed mainly by paired helical filaments (PHFs) which are formed by hyperphosphorylated and aggregated tau (PHF-tau) [13]. On the other hand, potentially pathogenic forms of tau are triggered by diverse environmental factors producing tissue damage such as stroke, head injury, stress, exposure to aluminium, traumatic brain injury, starvation, and cold water stress [15].

Among the earliest detectable features in neurodegenerative diseases are the loss of neuronal synapses and dying-back of axons, which appear to be accompanied by a decay of the intracellular transport and correlates with the incipient loss of memory and brain functions [14]. Several triggering events such as oxidative stress, inflammatory cytokines, lack of growth factors, or the toxic amyloid-β (Aβ) peptide have been implicated in axonal and/or neuronal decay [14]. More recently, axonal injury and microglia activation have been observed in an experimental HSE model, suggesting the participation of HSV-1 in neuronal damage [16].

In a recent study, we infected mice neuronal cultures with HSV-1 to study changes in cytoskeleton dynamics and neurodegenerative damage. Our results showed that HSV-1 neuronal infection triggered hyperphosphorylation of tau. The first sites to be phosphorylated are epitopes S^{202} and T^{205} (Tau1 antibody) and then to epitopes S^{396} and S^{404} (PHF1 antibody). Infection produced marked neurite damage and an important reduction of neuronal viability (Figs. 1 and 2) [17]. These results suggest that HSV-1 infection could lead to neuronal cytoskeleton disruption and neurodegenerative processes, and reinforces the idea that recurrent viral reactivation episodes in vivo could trigger neurodegenerative processes.

Another interesting aspect is that Aβ fibrils have been demonstrated to stimulate in vitro infection by enveloped viruses, including HSV [18]. In addition, glycoprotein B of HSV-1, whose internal sequence shows homology to the C-terminal region of Aβ, has been demonstrated to promote fibril formation in vitro [19]. Thus, a close association between Aβ deposition and HSV-1 reactivation in the human brain appears to exist.

3   Neuroinflammation

Brain inflammation due to infection, aging, and other deleterious processes is associated with activation of the local innate immune system. In contrast to other tissues, the CNS is essentially devoid of MHC expression and shielded from antibodies by the blood–brain

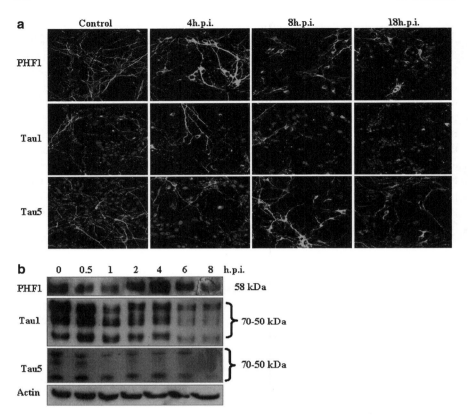

Fig. 1 HSV-1 infection results in hyperphosphorylation of tau protein. **a** Primary cortical neurons infected with HSV-1 (moi 10) for 4, 8, and 18 h and untreated controls were stained with specific antibodies for PHF1, Tau1, and Tau5. Nuclei were counterstained with propidium iodide. The results are representative of three separate experiments. Magnification 100×. **b** Western blot analyses of PHF1, Tau1, Tau5, and actin in primary cortical neurons treated with HSV-1 (moi 10) at 0, 0.5, 1, 2, 4, 6, and 8 h post-infection (*hpi*). Blots shown are representative of three separate experiments. Reprinted from the *Journal of Alzheimer's Disease* (in press), Copyright (2008), with permission from IOS Press (*See Color Plates*)

barrier. Therefore, a rapid local innate immune response by resident cells is required to effectively fight infectious agents. These microglial cells detect pathogen-associated molecular patterns and also toxic cell debris, such as Aβ fibrils and other aggregated proteins, through a variety of pattern-recognition receptors including complement, Toll-like receptors (TLRs), scavenger receptors, among many others [20].

Increasing evidence indicates that TLRs play a major role in several inflammatory CNS pathologies. TLR-mediated intracellular signalling pathways converge to activate nuclear factor-κB (NF-κB) and c-Jun N-terminal kinases, which induce the transcription of a series of cytokine/chemokine genes that are involved in the initiation or regulation of the inflammatory response. In the CNS, TLRs are expressed

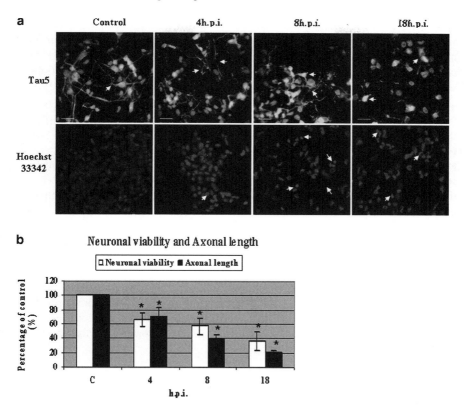

Fig. 2 Neuronal cultures infected with HSV-1 exhibit axonal disruption. **a** Neurons infected with HSV-1 (moi 10) at 4, 8, and 18 h post-infection (*hpi*) and untreated controls were stained with a specific total tau antibody, Tau5, and the nuclei were counterstained with propidium iodide (*upper row*) or with Hoechst 33342 (*lower row*). **b** Neuronal viability and axonal length in HSV-1-infected cells were expressed at different times post-infection as a percentage of the values obtained in uninfected control. **c** Data represent mean ± SEM for three independent experiments ($^*p < 0.05$ compared to untreated control). Reprinted from the *Journal of Alzheimer's Disease* (in press), Copyright (2008), with permission from IOS Press (*See Color Plates*)

predominantly by astrocytes, the most abundant glial cell population, and microglial cells, which are considered the 'CNS professional macrophages' [21]. Recent studies suggest that TLRs, mainly TLR2 and TLR9, are important players during HSV infection, initiating a signalling cascade leading to NF-κB activation and proinflammatory cytokine secretion [22–24]. Although TLR signalling is essential in anti-microbial defense, it may also constitute a 'double-edged sword' that leads to injury of CNS tissue caused by an excessive inflammatory response [25]. Moreover, it has been suggested that TLR2 expression would not be protective, but rather associated with lethal viral encephalitis on HSV-1 infection in mice [22]. After the inflammatory burst caused by HSV-1 infection, microglial cells undergo apoptosis, and both processes have been shown to be induced through TLR2 signalling [26]. TLR2-mediated signalling has also been involved in neuroinflammation and neuronal damage caused by bacterial meningitis [27]. However, it remains controversial whether microglial cells have beneficial or detrimental functions in various neuropathological conditions [28–30]. It has been proposed that chronic activation of microglia may cause neuronal damage through the persistent release of potentially cytotoxic molecules (e.g., proinflammatory cytokines, reactive oxygen intermediates, proteinases, and complement proteins) [30]. But also an acute phase protein, such as α-1-anti-chymotripsin, has been shown to be over-expressed in astrocytes of AD patients and to induce tau hyperphosphorylation in neurons [31]. In the context of tau hypothesis, a solid set of discoveries has strengthened the idea that neuroinflammation is responsible for an abnormal secretion of proinflammatory cytokines that trigger signalling pathways that activate brain tau hyperphosphorylation [30]. On the other hand, microglial cells and TLRs display important roles in brain development and neuronal repair [32]. Several other reviews dealing with TLR receptors and their roles in HSV infection, CNS inflammation, and homeostasis have been published recently [33–35].

All the results mentioned above suggest that repeated reactivation of HSV-1 in neurons would likely promote neurodegeneration in brain. Therefore, preventive and therapeutic measures against HSV-1 or other chronic inflammatory infections should be applied not only to diminish the incidence of sporadic but also to chronic neurodegenerative diseases.

Acknowledgements The authors would like to thank Ilona Concha, Ph.D. for her comments. The study referred was financially supported by Grant 14060/05 from Fundación Andes, S-2007-62 from DID-UACH and MECESUP AUS0107.

References

1. Dobson CB, Wozniak MA, Itzhaki RF (2003) Do infectious agents play a role in dementia. Trends Microbiol 11:312–317
2. Whitley RJ, Gnann JW (2002) Viral encephalitis: familiar infections and emerging pathogens. Lancet 359:507–513
3. Wozniak MA, Shipley SJ, Combrinck M et al. (2005) Productive herpes simplex virus in brain of elderly normal subjects and Alzheimer's disease patients. J Med Virol 75:300–306
4. Itzhaki R (2004) Herpes simplex virus type 1, apolipoprotein E and Alzheimer' disease. Herpes 2:77A–82A
5. Ball, MJ. 1982. Limbic predilection in Alzheimer dementia: is reactivated herpesvirus involved? Can J Neurol Sci 9:303–306
6. Grant WB, Campbell A, Itzhaki RF et al. (2002) The significance of environmental factors in the ethiology of Alzheimer's disease. J Alzheimers Dis 4:179–189
7. Itzhaki RF, Lin WR, Shang D et al. (1997) Herpes simplex virus type 1 in brain and risk of Alzheimer disease. Lancet 349:241–244
8. Jamieson GA. Maitland NJ, Wilcock CK et al. (1992) Herpes simplex virus type 1 DNA is present in specific regions of brain from aged people with and without senile dementia of the Alzheimer type. J Pathol 167:365–368

9. Bertrand P, Guillaume D, Hellauer K et al. (1993) Distribution of Herpes simplex virus type 1 DNA in selected areas of normal and Alzheimer's disease brains: a PCR study. Neurodegeneration 2:201–208

10. Hill JM, Gebhardt BM, Azcuy AM et al. (2005) Can a herpes simplex virus type 1 neuroinvasive score be correlated to other risk factors in Alzheimer's disease? Med Hypotheses 64:320–327

11. Mori I, Kimura Y, Naiki H et al. (2004) Reactivation of HSV-1 in the brain of patients with familial Alzheimer's disease. J Med Virol 73(4):605–611

12. Goedert M, Spillantini MG, Jakes R et al. (1989) Multiple isoforms of human microtubule-associated protein Tau: sequences and localization in neurofibrillary tangles of Alzheimer's disease. Neuron 3:519–526

13. Goedert M, Spillantini MG (2006) A century of Alzheimer's disease. Science 314:777–781

14. Mandelkow EM, Stamer K, Vogel R et al. (2003) Clogging of axons by tau, inhibition of axonal traffic and starvation of synapses. Neurobiol Aging 24:1079–1085

15. Okawa Y, Ishiguro K, Fujita SC (2003) Stress-induced hyperphosphorylation of tau in the mouse brain. FEBS Lett, 535:183–189

16. Mori I, Goshima F, Mizuno T et al. (2005) Axonal injury in experimental herpes simplex encephalitis. Brain Res 1057:186–190

17. Zambrano A, Solis L, Salvadores N et al. (2008) Neuronal cytoskeletal dynamic modification and neurodegeneration induced by infection with herpes simplex virus type 1. J Alzheimers Dis 14:in press

18. Wojtowicz WM, Farzan M, Joyal JL et al. (2002) Stimulation of enveloped virus infection by β-amyloid fibrils. J Biol Chem 277:35019–35024

19. Cribbs DH, Azizeh BY, Cotman CW et al. (2000) Fibril formation and neurotoxicity by a herpes simplex virus glycoprotein B fragment with homology to the Alzheimer's Aβ peptide. Biochemistry 39:5988–5994

20. Griffiths M, Neal JW, Gasque P (2007) Innate immunity and protective neuroinflammation: new emphasis on the role of neuroimmune regulatory proteins. Int Rev Neurobiol 82:29–55

21. Farina C, Aloisi F, Meinl E (2007) Astrocytes are active players in cerebral innate immunity. Trends Immunol 28:138–145

22. Kurt-Jones EA, Chan M, Zhou S et al. (2004) Herpes simplex virus 1 interaction with Toll-like receptor 2 contributes to lethal encephalitis. Proc Natl Acad Sci USA 101:1315–1320

23. Aravalli RN, Hu S, Rowen TN et al. (2005) Cutting Edge: TLR2-mediated proinflammatory cytokine and chemokine production by microglial cells in response to herpes simplex virus. J Immunol 175:4189–4193

24. Mansur DS, Kroon EG, Nogueira ML et al. (2005) Lethal encephalitis in myeloid differentiation factor 88-deficient mice infected with Herpes simplex virus 1. Am J Pathol 166:1419–1426

25. Crack PJ, Bray PJ (2007) Toll-like receptors in the brain and their potential roles in neuropathology. Immunol Cell Biol 85:476–480

26. Aravalli RN, Hu S, Lokensgard JR (2007) Toll-like receptor 2 signaling is a mediator of apoptosis in herpes simplex virus-infected microglia. J Neuroinflammation 4:11–17

27. Hoffmann O, Braun JS, Becker D et al. (2007) TLR2 mediates neuroinflammation and neuronal damage. J Immunol 178:6476–6481

28. Marques CP, Hu S, Sheng W et al. (2006) Microglial cells initiate vigorous yet non-protective immune responses during HSV-1 brain infection. Virus Res 121:1–10

29. Glezer I, Simard AR, Rivest S (2007) Neuroprotective role of the innate immune system by microglia. Neuroscience 147:867–883

30. Rojo LE, Fernández JA, Maccioni AA et al. (2008) Neuroinflammation: implications for the pathogenesis and molecular diagnosis of Alzheimer's disease. Arch Med Res 39:1–16

31. Padmanabhan J, Levy M, Dickson DW et al. (2006) Alpha I-antichymotrypsin, an inflammatory protein overexpressed in Alzheimer's disease brain, induces tau phosphorylation in neurons. Brain 129:3020–3034

32. Larsen PH, Holm TH, Owens T (2007) Toll-like receptors in brain development and homeostasis. Sci STKE 402:47

33. Finberg RW, Knipe DM, Kurt-Jones EA (2005) Herpes simplex virus and Toll-like receptors. Viral Immunol 18:457–465
34. Herbst-Kralovetz M, Pyles R (2006) Toll-like receptors, innate immunity and HSV pathogenesis. Herpes 13:37–41
35. Kielian T (2006) Toll-like receptors in central nervous system glial inflammation and homeostasis. J Neurosci Res 83:711–730

General Aspects of AD Pathogenesis

Selective Cerebrocortical Regional, Laminar, Modular and Cellular Vulnerability and Sparing in Alzheimer's Disease: Unexploited Clues to Pathogenesis, Pathophysiology, Molecular- and Systems-Level Hypothesis Generation and Experimental Testing

Rodrigo O. Kuljis

Abstract Virtually all current hypotheses on the pathogenesis and pathophysiology of Alzheimer's disease rely on an unvoiced "amorphous" concept of the brain that essentially ignores its highly complex organization at the systems neuroscience level. This is especially true for the cerebral cortex, which happens to be the main target of the disorder and arguably the most complex structure of the entire brain. Here I review increasing evidence that the involvement of the cortex – while abundant – is not diffuse, random, or chaotic. In fact, the highly stereotyped patterns of the three-dimensional involvement of the cerebral cortex indicate that the pathobiological process targets highly selected cells and both anatomically and functionally unique multicellular arrays, while closely situated elements appear considerably resistant to the disease process. This remarkable dichotomy seems to apply pancortically and has essentially escaped recognition by most students of the disorder. Not surprisingly, there is no explanation for the selective involvement versus sparing of circuitry that is immediately adjacent of one another, and this notion seems conspicuously absent from virtually all models of the disorder. In fact, none have so far ever addressed the now highly probable central role of the modular organization of this region in the emerging pattern of vulnerability versus resistance to the disease process. This situation calls for an integration of at least the molecular and the systems neuroscience approaches to formulate new hypotheses on the pathogenesis and pathophysiology of Alzheimer's disease, in order to enter a new stage in the elucidation of the disorder that accounts better for the factors that make certain neuronal assemblies more vulnerable – while others seem to be distinctly resistant – and precisely how this helps accounts for the clinical manifestations of the disease. Recent observations in animal models of

R.O. Kuljis
Cognitive Sciences Division
Brain-Mind Project, Inc., and Encephalogistics,
Inc., Miami, FL 33131, USA.

R.B. Maccioni and G. Perry (eds.) *Current Hypotheses and Research Milestones*   191
in Alzheimer's Disease. DOI: 10.1007/978-0-387-87995-6_16,
© Springer Science + Business Media, LLC 2009

some of the factors that may influence the selectivity of the disease process in the cerebral cortex open the possibility of testing novel hypotheses experimentally, and thus eventually extending the results to translational efforts aimed at new, more effective treatments, early diagnosis, and prevention.

1 Introduction

Alzheimer's disease usually begins with a progressive deterioration in memory, and over time additional symptoms accrue that reflect the involvement of other cognitive spheres. The latter symptoms may include aphasia, apraxia, agraphia, and alterations in visuospatial abilities [1, 2]. There are also a few well-documented cases in which, instead of memory loss, the initial manifestation involves a different cerebral cortex-mediated faculty, such as aphasia, "posterior cortical atrophy," or a right parietal syndrome [3–11].

Here, I review and reconceptualize the presumed mechanisms responsible for the symptoms of Alzheimer's disease, attempting an analysis of its anatomical and histopathological substrate in order to deduce some of the pathophysiological processes responsible for the main clinical manifestations. As elaborated below, the bulk of the experimental evidence suggests very strongly that this is a predominantly cortical disorder, first recognized by the prescient statement of Alois Alzheimer himself [12] that this is a disorder of the cerebral cortex. Within the general framework of cortical involvement, however, there is a clear predilection not only for certain regions of the cortex but also for certain layers of this structure and, as is well known, for certain types of neurons: pyramidal cells. After having reviewed briefly the latter aspects, we end by focusing on a remarkable but essentially ignored topic: the mounting evidence that there is also a predilection for specific components of the hypothetical anatomical and functional unit of the cerebral cortex, the so-called cortical "module." This relatively new pathophysiological concept has key implications for the understanding of the disorder from the integrative and systems neuroscience perspective. Such a viewpoint may permit a new synthesis with the inevitably – and in fact, desirably – reductionist analysis of the disorder from the molecular perspective, since it allows for an integration of these two distinct levels of inquiry in order to better understand the basis for normal and disordered cognitive function and behavior. Along these lines, it would seem that – given the volume of information available – this novel paradigmatic view may eventually be applied to the understanding and treatment of many other disorders characterized by deterioration in cognition and behavior.

2 Alzheimer's Disease as a Cerebrocortical Process

Despite the potential variety of symptoms among patients – including unusual presentations and even those that are extremely rare such as aphasias and a right parietal syndrome – it seems reasonable to postulate that in virtually all cases, the symptoms reflect a cerebrocortical dysfunction [12–16]. The vast majority of studies also indicate that the main lesions situated in the brain – that is, senile plaques and neurofibrillary tangles – are situated predominantly in the cerebral cortex [12, 17]. Furthermore, even in those studies focusing on the subcortical involvement, such as those in the nucleus basalis of Meynert [18], in the *locus ceruleus* [19], or in the pulvinar nucleus of the thalamus [13], the pathophysiological interpretation of the findings is invariably made in terms of how they affect cortical function, and in no case contradict the fundamental importance attributed to the cortical lesions. Thus, it is postulated that lesions in the nucleus basalis of Meynert are important because they deprive the cerebral cortex of its cholinergic innervation [18], and that lesions in the pulvinar destroy cortico-thalamo-cortical circuits that interconnect this nucleus with a vast territory in the association cortex [13].

These notions, combined with the relative absence of lesions in the majority of subcortical regions – and especially in contrast with many other dementing conditions in which the pathology is situated predominantly (if not always exclusively) subcortically (e.g., Parkinson's disease dementia, Huntington's disease, and progressive supranuclear palsy) – support the contention that Alzheimer's disease is a cerebrocortical discon-nection disorder [8, 13]. This concept has allowed some of us to focus many studies in which we attempt to understand in detail the pathophysiology of the disorder analyzing in detail the topographic distribution of the lesions in the cerebral cortex [14–16, 20]. Such focus aims, among other objectives, to understand the mechanisms underlying the disorder in terms of the circuits that are presumably selectively affected, compared with circuits that tend to be relatively spared or resistant to the degenerative process. Such a perspective is rarely considered in parallel with that which tends to dominate inquiry in this field today: that is, the molecular factors involved in the pathogenesis and pathophysiology of the disease. Nevertheless, the analysis of the patterns of topographic distribution of lesions in the disorder, which is in its infancy, adds an important dimension – from the perspective of Systems Neuroscience – to the understanding of the pathophysiology of the condition. Such a perspective is undoubtedly essential in the brain, which is the organ with the broadest phenotypic diversity among its cells in the entire organism. Therefore, it is not reasonable to consider the role of any trophic or toxic agent – or any other molecular factor affecting the expression of the clinical phenotype – while ignoring the highly complex three-dimensional organization of the brain. In fact, assuming that most of the various cell types in the brain are exposed to the degenerative process, the bulk of the neuropathological studies demonstrate that specific types of cells are highly susceptible, whereas others are remarkably resistant to the disease (Fig. 1).

The great majority of studies reveal a considerably stereotyped pattern of involvement in the cortex: regions most affected include the entorhinal cortex (Brodmann's area 28), the perirhinal cortex (area 35), and the subjacent hippocampus and the temporopolar cortex (area 38) [14–16, 20]. Compared to these areas, the rest of the association cortex develops an intermediate density of lesions, whereas the primary cortices – both motor and sensory – develop the lowest density of neurofi-brillary tangles. It is important to emphasize, however, that this notion is predicated primarily on the density of neurofibrillary tangles, whose distribution is felt by

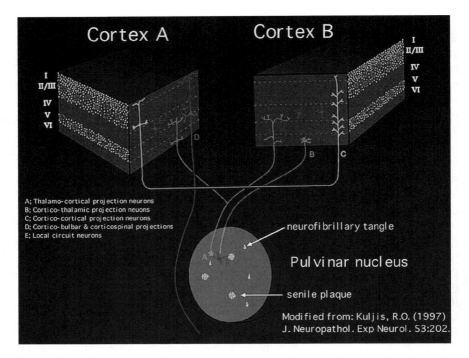

Fig. 1 Schematic rendering of the corticocortical and thalamocortical circuits affected by Alzheimer's disease. The *blocks* in the *upper part* of the figure represent parts of two different cortical areas. The *spheroid* in the *lower part* of the figure represents the pulvinar nucleus of the thalamus, one of the few in the entire thalamus that contain Alzheimer's disease lesions and is otherwise considered "unaffected." In all these representations, the *drop-like icons* signify neurofibrillary tangles, and the *semispheroidal yarn-like elements* signify senile plaques. *Roman numerals* represent cortical layers, and *regions with high densities of lesions* represent layers (V) or groups of layers (II/III) stereotypically affected in the disorder. With the exception of the pulvinar and a handful of very circumscribed regions, thalamocortical projections are essentially unaffected given the low density of lesions in layers IV and VI and in other subcortical regions. Similarly, the reciprocal corticothalamic projections (B) that originate primarily in layer VI also appear unaffected. In contrast, corticocortical connections that originate from medium-sized pyramidal neurons in layers II/III are distinctly and selectively affected given the high density of lesions in them. Other important elements in these circuits, which originate from layer V (labeled D in the figure), and some local circuit neurons (E, F) are also affected. Modified from Kuljis [13], with permission (*See Color Plates*)

many to correlate best with the level of cognitive deterioration [21–23], although it has been shown that primary sensory cortices have a high density of senile plaques [24] which is in fact higher than that in medial temporal regions that are widely perceived as severely affected [14]. Apart from this gross overall topographic pattern, there is also a distinct laminar predilection in the distribution of the lesions (Fig. 1). This pattern applies both to allocortical and to iso(neo)cortical regions, and consists generally in a higher density of lesions in layers from which corticocortical projections originate (layers II/III in the neocortex), as compared with layers that receive projections from subcortical regions (I and IV) or layers from where feedback cortico-subcortical projections originate (VI). Such a pattern permits postulating that the disposition of the lesions explains the cortical dysfunction in terms of a disconnection resulting from the disruption of corticocortical connectivity [24, 25].

3   Modular Organization of the Cerebral Cortex and Its Relevance to Alzheimer's Disease

Apart from the selective distribution of lesions in different regions and areas of the cerebral cortex, and, especially, in different layers, there is a third level or category of organization in which their disposition is also uneven but apparently not randomly so: the modular level of organization. This domain of cortical organization has been increasingly recognized as important because it represents both an anatomical and a functional unit essential to understanding both normal and disordered function, although neglected of recent in favor of molecular studies. This domain of organization spans the thickness of the cerebral cortex, that is, all of its layers, including those laden with lesions (both senile plaques and neurofibrillary tangles), and few investigators seem to have appreciated this pattern so evident in certain regions of the cortex [24, 26–30].

Perhaps the first indication that the cerebral cortex is made of iterated macrocellular arrays was reported by Arnold, who designated what would later be named the entorhinal cortex as the "glomerular substance" [31]. This observation was verified subsequently by several neuroanatomists, who recognized in its surface a series of iterated elevations surrounded by "valleys" [32, 33] – unique in the entire cerebral cortex (Fig. 2) – that

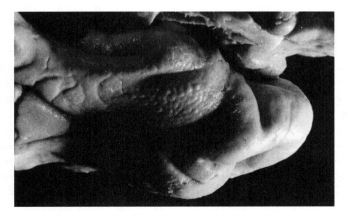

Fig. 2 Modular organization of the entorhinal cortex. Photomacrograph of the medial aspect of the temporal lobe where the "glomeruli" or "warts" that decorate the surface of the entorhinal cortex are apparent and demarcate the extent of this area. From the author's collection (*See Color Plates*)

were rebaptized as "warts" by Klinger [34]. However, since the concept of modular organization did not exist then, it was not recognized until much later that this feature of the entorhinal cortex reflects at the macroscopic level its modular organization as evident by even cursory examination of its cytoarchitecture. In fact, the realization of the importance of this observation occurred over more than a century after the discovery of the "glomerular substance," after vaguely homologous multicellular aggregates were discovered in neocortical areas, which I summarize below.

Since the pioneering studies of Santiago Ramón y Cajal on the organization of the cerebral cortex, it has been apparent that many of the then probable connections among neurons are predominantly traversing the thickness of the cerebral cortex [35]. However, the first indications of a functional "columnar" organization were eventually confirmed and extended by Rafael Lorente de Nó (1902–1990), one of his pupils, when he discovered the now-named "barrels" (his "glomeruli") in what was subsequently identified as a part of the somatosensory cortex of the mouse [36]. These barrels consist of semicylindrical aggregates of neurons in layer IV, which correspond merely to the cytoarchitectonically obvious part of anatomical and functional units composed of a large array of neurons that span the thickness of this region of the cortex [37]. The subsequent studies of Mountcastle in the somatosensory cortex of nonhuman primates – which, nevertheless, lack barrels – provided the best evidence that this region is made of roughly cylindrical "units" made of columnar arrays of neurons spanning the cortical thickness which share physiological properties and thus appear to help perform unique "computations," "operations," or "permutations" of information that are distinct from those performed by neighboring columns [38], reviewed by Kuljis and Rakic [39, 40]. Later, and mainly as the result of studies in the primary visual (striate) cortex in primates, it was found that this area is made also by functional units of roughly columnar disposition. Furthermore, it was also discovered that there are functionally and anatomically distinct types of "columns" adjacent to each other and that an aggregate of columns presumably "analyzing" the various attributes of visual stimuli presented in a determined portion of the visual field is thought to constitute "modules" or "hypercolumns" (Fig. 3) of which the visual cortex is built [41]. Theoretically, then, it is proposed that the entire cerebral cortex is made of a mosaic of modules whose internal organization and interconnections to the various sense organs – as well as to other regions of the cortex – differs among regions, thus endowing each set of modules (and, hence, cortical areas) with their more or less unique functional properties [38–43]. However, although this principle of organization is widely accepted, it is not evident in areas that lack anatomically obvious modular cytoarchitectonic or chemoarchitectonic organization, or, lacking these, a means to put such an organization in evidence by electrophysiological methodology [38–40]. Therefore, this theoretical principle of pancortical organization is not completely devoid of detractors [44, 45]. Fortunately, the theoretical concept of the modular organization of the cerebral cortex is helpful in the formulation of hypotheses about the mode of involvement of this region in neurodegenerative disorders, and the experimental results obtained so far provide novel evidence for both the modular organization of the cortex and for its differential involvement in Alzheimer's disease, discussed below.

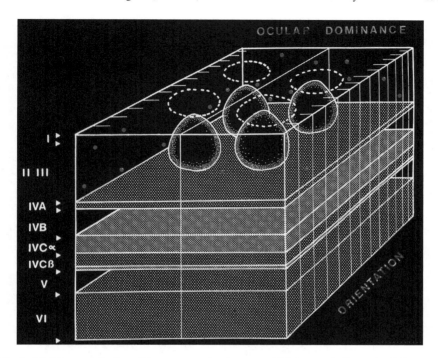

Fig. 3 Modular organization of the macaque visual cortex, which is very similar to that in humans. Schematic representation of a segment of this area of the cortex as a cube in which the *upper surface* corresponds to the piamater and the *lower surface* corresponds to the interface between the gray and white matter. *Roman numerals* depict cortical layers and hatching regions with a high cytochrome oxidase (*CO*) activity. The *ovoid objects* in layers II/III correspond to the "patches" rich in CO that contain neurons that respond predominantly to wavelength (color) of visual stimuli as opposed to those responsive mainly to orientation in the interpatch spaces of the same layers. This wavelength versus orientation selectivity spans the thickness of the cortex in semicylindrical arrangements that form the basis of the "hypercolumns" postulated by Hubel and Wiesel [41] according to which this cortex consists of an array of hypercolumns. Modified from Kuljis and Rakic [39], with permission (*See Color Plates*)

4 Evidence for the Modular Distribution of Alzheimer's Disease Lesions

As summarized above, it was known since the middle of the nineteenth century, and especially since the middle of the twentieth century that layer II of the entorhinal cortex is discontinuous, forming "islets" which – on the surface – develop elevations or hillocks known as "warts" [29–34]. This is in contrast with the remainder of the cortex which has a layer II or equivalent that is uninterrupted. Layer II islets in the entorhinal cortex thus serve as putative indicators – even after death – of the functional units that can be designated as "modules" (Fig. 2). In the visual cortex, histochemical methods such as an enzymatic reaction for cytochrome oxidase allow visualization of the "spots," "blots," "patches," or "puffs" situated in layers II/III, which demarcate the centers of the

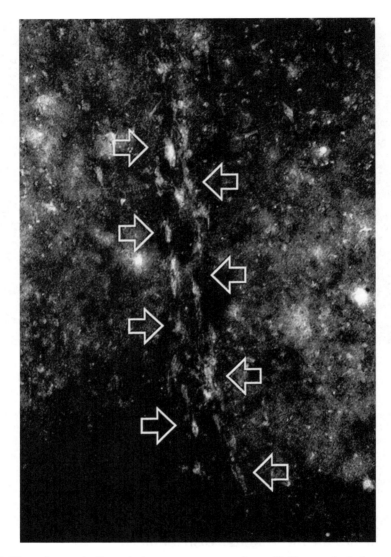

Fig. 4 Photomicrograph of a semicolumnar arrangement of neurofibrillary tangles in the perirhinal cortex. Thioflavin S epifluorescence in a patient with histopathologically confirmed Alzheimer's disease. The pial surface is toward the bottom and the pseudocolumnar array of the lesions spans the thickness of this cortical region. From the author's collection

hypercolumns or modules [39–42, 46, 47] that presumably compose this area (Fig. 4). Thus, cytochrome oxidase histochemistry and other histochemical methods have been useful to begin to determine whether Alzheimer's disease lesions are situated preferentially in the modules, or in parts of the modules, revealing a level of selectivity heretofore unknown in the histopathological process [24, 26, 30].

For example, in the entorhinal cortex, neurofibrillary tangles are situated in virtually all the neurons in the islets of layer II [15, 25, 30]. This has been interpreted as a selective involvement of the modules making this region, and is perhaps the first anatomically evident indication of selectively disposed pathology along the "tangential" (as opposed to the "vertical," i.e., translaminar) domain of the cortex. Although this is an important and compelling observation in the entorhinal cortex, there is an equally important concern that remains unresolved: Since layer II of the entorhinal cortex is discontinuous, and given that its histopathological involvement occurs early in the disease process and is rarely if ever evaluated until the patient dies in advanced stages of the disorder, why is one to expect other than the lesion distribution reflects merely the anatomical disposition of neurons in this layer? In other words, Alzheimer's disease may merely replace one marker of its modular organization (i.e., islets of neurons) with another ("tombstone" neurofibrillary tangles that afflict all the pyramidal neurons in this layer, after the neurons that contained them died). From this perspective, the presence of pathology with modular-like features may be completely predictable, and may be postulated as a mere reflection in the histopathology of its formerly normal organization. It could thus be argued that the module-like disposition of the neurofibrillary tangles in the entorhinal cortex is only a curious anomaly devoid of functional significance for the rest of the cerebral cortex.

The latter concern has begun to be addressed by the discovery of discontinuous and iterated pathology in other areas of the cortex devoid of anatomically evident modules (i.e., discontinuously disposed neurons) in normal individuals (Fig. 5). For example, Akiyama et al. observed a column-like arrangement of amyloid-β deposits in the cerebral cortex of patients with Alzheimer's disease [28]. These deposits 200–600 μm in width in Brodmann's areas 9, 39, 21, 4, 3, 1, 17, and 18 are strongly reminiscent of the columns postulated by Lorente de Nó [36], Szenthágothai [43], and all subsequent authors, but unfortunately this disposition does not automatically prove that they do indeed correspond to such columns. This is because proving the functional nature of such columnar arrangement would require histochemical labeling of markers of such function, which were not attempted and unfortunately do not exist in most of these areas, and because it is not possible to perform electrophysiological verification of columns in postmortem tissue. The observation is very important, however, since these column-like deposits occur in areas known to contain a modular organization (Brodmann's areas 18, 17, 9, 4, 3, and 1) as well as at least two others widely suspected to have this mode of organization, but in which it has not been rigorously demonstrated to date.

The latter difficulty has begun to be addressed with the discovery of discontinuous and iterative pathology in convincingly modularly organized regions of cortex, which is highly suggestive of a selective modular involvement by the pathology in Alzheimer's disease [24, 27]. These include the perirhinal and primary visual (striate) cortices, both of which have continuous layers, that is, lacking the fragmentation in layer II of the entorhinal cortex which probably imparts a compartmentalized distribution to the neurofibrillary tangles that develop in this latter layer, and thus permitting a potential test to the hypothesis that such module-like distribution reflects a modular bias on Alzheimer's disease pathology. Therefore, it is quite compelling that the perirhinal (Fig. 4) and the primary visual cortices (Fig. 5) – which have a completely different

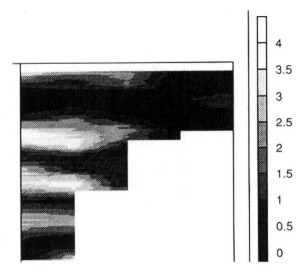

Fig. 5 Two-dimensional representation of the density distribution of senile plaques in layers II/III of the primary visual cortex in a patient with Alzheimer's disease. The density of these lesions is depicted as if the layers were viewed tangentially from the cortical surface after computer-assisted unfolding and stereological analysis. Progressively *lighter regions* represent high densities, whereas progressively *darker lesions* represent lower densities of lesions. It is evident that the density of senile plaques alternates in the horizontal plane of layers II/III, which is consistent with the hypothesis that they are contained more or less selectively in components of the iterated modules or hypercolumns that constitute the "building blocks" of this area of the cerebral cortex. Modified from Kuljis and Tikoo [24], with permission

structure, phylogenetic origin, and function, but nevertheless can be hypothesized to be organized into modules – develop an inhomogeneous distribution of lesions across layers suggestive of the iterated compromise of elements of such modules (see below). This alleviates substantially the difficulties of postulating such hypothesis based solely on observations in the entorhinal cortex.

The column-like lesion aggregates in the perirhinal cortex (Fig. 4) likely represent part of iterated circuits that constitute its modules, which, for reasons that remain to be determined, appear to be vulnerable to the disease process earlier than immediately neighboring circuits. A similar interpretation can be made of a more elaborate analysis in the striate cortex, in which we used computer-assisted strategies to "unfold" this region and map the senile plaques in layers II/III to a "midcortical plane" to render a representation of their distribution as if one were observing them from the pial surface on a "flat mount." Such two-dimensional maps have the advantage of permitting visualizing the distribution of the lesions in a way that is impossible by cursory examination of individual sections, permitting also a comparison of their distribution with well-known features of the organization of the striate cortex, such as the eye dominance columns (Fig. 5). Such analysis reveals that the senile plaques in layers II/III are disposed in a pseudocolumnar arrangement in the striate cortex that is highly reminiscent of eye dominance columns, providing the best evidence so far that at least some of the lesions in the cerebral cortex are situated selectively within the modules that compose the areas in which they occur [24, 27]. The pattern in the visual cortex is geometrically distinct from that in the entorhinal and perirhinal cortices, and in no case interpretable as due to polysynaptic interconnections among these three considerably different and – comparing the parahippocampal cortices with the visual cortex – substantially far away cortices. Therefore, this relatively newly elucidated pattern in the visual cortex is an independent confirmation, which is not predictable from cytoarchitectonics, of the hypothesis that Alzheimer's disease targets selectively iterated cerebrocortical circuits that are distinct if not unique among most cerebral cortices, and that constitute components of the hypothetical "modules" that correspond to macroneuronal complexes essential for cortical structural organization and function [24].

5 Conclusions

Taken together, the findings summarized above support the prevalent concept that the cerebrocortical involvement in Alzheimer's disease is extensive and abundant, and in contrast with the rather sparse involvement in subcortical regions. However, in contrast with the seemingly prevalent dogma, our observations and those of several others indicate that the cortical involvement – while abundant – is not diffuse, random, or chaotic. In fact, the highly stereotyped patterns of the three-dimensional involvement of the cerebral cortex indicate highly selected targets on the part of the pathobiological process, including groups of increasingly more easily identifiable neurons that appear considerably resistant to the disease process, although they are in the immediate vicinity of severely affected neurons. This dichotomy has essentially escaped recognition by most students of the disorder, and seems conspicuously absent from virtually all models of Alzheimer's disease that ignore completely the highly complex organization and function of the cerebral cortex – the main target of the disease process – and have so far never addressed the now highly probable central role of the modular organization of this region. This situation calls for an integration of the molecular and Systems Neuroscience approaches to reformulate new hypotheses on Alzheimer's disease pathogenesis and pathophysiology, in order to enter a new stage in the elucidation of the disorder that accounts better for the factors that appear to make certain neuronal assemblies more vulnerable, whereas others are distinctly resistant, and precisely how this helps account for the clinical manifestations of the disease. It is quite obvious that such improved understanding could be readily applicable to improved means for presymptomatic diagnosis (i.e., imaging modularly distributed changes as predicted by Damasio et al. [48]), as well as in the design and implementation of novel preventative and palliative strategies inspired by a mechanistic understanding of selective sparing versus degeneration of cerebrocortical circuitry.

Newly available and heretofore unexploited opportunities to advance these goals exist now in the form of both radiation-induced [49] and genetically engineered animals that can serve as models of some aspects of Alzheimer's disease pathology,

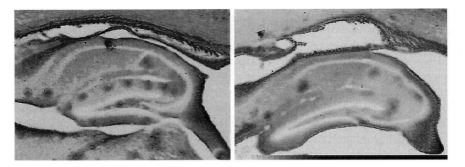

Fig. 6 Photomicrographs of amyloid-containing-like lesions in the hippocampus of transgenic mice overexpressing normal human-α 1 antichymotrypsin. Gallyas' method for amyloid deposits (which is different from other methods from the same author to demonstrate myelin and cytoskeletal lesions). Note that the lesions in the hippocampus tend to have a selective laminar disposition, reflecting a propensity to affect certain elements of this structure, while sparing others in a manner similar to that in patients with Alzheimer's disease. Modified from Kuljis et al. [50] (*See Color Plates*)

which exhibit suggestions of selective laminarly and modularly distributed "lesions" (Fig. 6). It is remarkable indeed that some rather recent experimental manipulations result in changes in the disposition of lesion-like changes [50, 51] that suggest that it may now be possible to dissect some of the molecular factors that target specific neuronal arrays in the cortex, while sparing others, and how this may change as a result of altering the balance of vulnerability versus protective factors involved in a specific disease model. The fact that all of this occurs in existing experimental animals in which the modular organization of the cerebral cortex is readily assessable both anatomically as well as physiologically – with a vast array of exquisite newly available methodology – provides tantalizing opportunities that few seem prepared to consider, and much less to exploit.

References

1. Cummings JL, Benson DF (1992) Dementia: A Clinical Approach. Boston: Butterworth-Heinemann
2. Katzman R (1986) Alzheimer's disease. New Engl J Med 314: 964–973
3. Ardila A, Rosselli M, Arvizu L, Kuljis RO (1997) Alexia and agraphia in posterior cortical atrophy. Neuropsychiatry Neuropsychol Behav Neurol 10: 52–59
4. Berthier ML, Leiguarda R, Starkstein SE, Sevlever G, Taratuto AL (1991) Alzheimer's disease with posterior cortical atrophy. Neurol Neurosurg Psychiatry 54: 1110–1111
5. Méndez MF, Zander BA (1991) Dementia presenting with aphasia: clinical characteristics. Neurol Neurosurg Psychiatry 54: 542–545
6. Levine DN, Lee JM, Fisher CM (1993) The visual variant of Alzheimer's disease. Neurology 43: 305–313
7. Jagust WJ, Davies P, Tiller-Borcich JK, Reed BR (1990) Focal Alzheimer's disease. Neurology 40: 14–19

8. Hof PR, Bouras C, Constantinidis J, Morrison JH (1989) Balint's syndrome in Alzheimer's disease: specific disruption of the occipito-parietal visual pathway. Brain Res 493: 368–375
9. Braak H, Braak E, Kalus P (1989) Alzheimer's disease: areal and laminar pathology in the occipital isocortex. Acta Neuropathol 77: 494–506
10. Gordon B, Selnes O (1984) Progressive aphasia "without dementia": evidence of more widespread involvement. Neurology 34 (Suppl.): 102
11. Crystal HA, Horoupian DS, Katzman R, Jotkowitz S (1982) Biopsy proven Alzheimer's disease presenting as a right parietal lobe syndrome. Ann Neurol 12: 186–188
12. Alzheimer A (1907) A singular disorder that affects the cerebral cortex. In: Hochberg, CN, Hochberg, FH, 1977. Neurologic Classics in Modern Translation. New York: Hafner Press, pp. 41–43
13. Kuljis RO (1994) Lesions in the pulvinar in patients with Alzheimer's disease. J Neuropathol Exp Neurol 53: 202–211
14. Arnold SE, Hyman BT, Flory J, Damasio AR, Van Hoesen GW (1991) The topographical and neuroanatomical distribution of neurofibrillary tangles and neuritic plaques in the cerebral cortex of patients with Alzheimer's disease. Cereb Cortex 1: 103–116
15. Braak H, Braak E (1991) Neuropathological stageing of Alzheimer-related changes. Acta Neuropathol 82: 239–259
16. Brun A, Englund E (1981) Regional pattern of degeneration in Alzheimer's disease: neuronal loss and histopathological grading. Histopathol 5: 549–564
17. Tomlinson BE (1992) Ageing and the dementias. In: Hume Adams J and Duchen LW (eds.), Greenfield's Neuropathology. New York: Oxford, pp. 1284–1410
18. Whitehouse PJ, Price DL, Clark AW, Coyle JT, Delong MR (1981) Alzheimer disease: evidence for selective loss of cholinergic neurons in the nucleus basalis. Ann Neurol 10: 122–126
19. Tomlinson BE, Irving D, Blessed G (1981) Cell loss in the locus ceruleus in senile dementia of Alzheimer type. J Neurol Sci 49: 213–219
20. Brun A, Gustafson L (1976) Distribution of cerebral degeneration in Alzheimer's disease. Arch Psychiatr Nervenkr 223: 15–33
21. Arriagada PV, Growdon JH, Hedley-Whyte ET, Hyman BT (1992) Neurofibrillary tangles but not senile plaques parallel duration and severity of Alzheimer's disease. Neurology 42: 631–639
22. Katzman R, Terry RD, De Teresa R et al. (1988) Clinical, pathological and neurochemical changes in dementia; a subgroup with preserved mental status and numerous cortical plaques. Ann Neurol 23: 138–144
23. McKee AC, Kosik KS, Kowall NW (1991) Neuritic pathology and dementia in Alzheimer's disease. Ann Neurol 30: 156–165
24. Kuljis RO, Tikoo RK (1997) Discontinuous distribution of lesions in striate cortex hypercolumns in Alzheimer's disease. Vision Res 37: 3573–3591
25. Hyman BT, Damasio AR, Van Hoesen GW, Barnes CL (1984) Alzheimer's disease: cell specific pathology isolates the hippocampal formation. Science 225: 1168–1170
26. Kuljis RO (1997) Modular corticocerebral pathology in Alzheimer's disease. In: Mangone CA, Allegri RF, Arizaga RL and Ollari JA (eds.), Dementia: A Multidisciplinary Approach. Editorial Sagitario: Buenos Aires, Argentina, pp. 143–155
27. Kuljis RO, Tikoo RK (1994) Tangentially selective distribution of amyloid-containing plaques in the striate cortex of Alzheimer's disease. Neurology 44 (Suppl.): A371 (959S)
28. Akiyama H, Yamada T, McGeer PL, Kawamata T, Tooyama I, Ishii T (1993) Columnar arrangement of β-amyloid protein deposits in the cerebral cortex of patients with Alzheimer's disease. Acta Neuropathol 85: 400–403
29. Van Hoesen GW, Solodkin A (1993) Some modular features of temporal cortex in humans as revealed by pathological changes in Alzheimer's disease. Cereb Cortex 3: 465–475
30. Solodkin A, Wu GF, Kuljis RO, Van Hoesen GW (1992) The perirhinal cortex (area 35) in man and its pathology in Alzheimer's disease. Soc Neurosci Abstr 18: 739 (307.8)
31. Arnold JC (1851) Handbuch der Anatomie des Menschen. Freiburg: Herder
32. Retzius G (1896) Das Menschenhirn. Studien in der Makroskopischen Morphologie. Stockholm: Norstedt & Söhne

33. von Economo C (1929) The Cytoarchitectonics of the Human Cerebral Cortex. New York: Oxford
34. Klinger J (1948) Denkschrifter der Schweizerischen Naturforeschenden Gesselschaft. Zurich: Bard
35. Ramón Cajal S (1899) Textura del Sistema Nervioso del Hombre y de los Vertebrados. Madrid: Moya
36. Lorente de Nó R (1922) La corteza cerebral del ratón. Trab. Lab. Invest. Biol. Univ. Madrid, 20, 41–78 (plus 25 plates)
37. Woolsey TA, Van der Loos H (1970) The structural organization of layer IV in the somatosensory region (SI) of mouse cerebral cortex. The description of a cortical barrel field composed of discrete cytoarchitectonic units. Brain Res 17: 205–242
38. Mountcastle VB (1957) Modality and topographic properties of single neurons of cat's sensory cortex. J Neurophysiol 20: 408–434
39. Kuljis RO, Rakic P (1989) Neuropeptide Y-containing neurons are situated outside cytochrome oxidase puffs in macaque visual cortex. Visual Neurosci 2: 57–62
40. Kuljis RO, Rakic P (1990) Hypercolumns in primate visual cortex can develop in the absence of cues from photoreceptors. Proc Natl Acad Sci USA 87: 5303–5306
41. Hubel DH, Wiesel TN (1977) Functional architecture of macaque visual cortex. Phil Trans R Soc Lond B 198: 1–59
42. Livingstone MS, Hubel DH (1982) Thalamic inputs to cytochrome oxidase-rich regions in monkey visual cortex. Proc Natl Acad Sci USA 79: 6098–6101
43. Szenthágothai J (1975) The "module-concept" in cerebral cortex architecture. Brain Res 95: 475–496
44. Purves D, Riddle DR, LaMantia AS (1992) Iterated patterns of brain circuitry (or how the cortex gets its spots). Trends Neurosci 15: 362–368
45. Swindale NV (1990) Is the cerebral cortex modular. Trends Neurosci 5: 345–350
46. Horton JC (1984) Cytochrome oxidase patches: a new cytoarchitectonic feature of monkey visual cortex. Phil Trans R Soc Lond B 304: 199–253
47. Horton JC, Hubel DH (1981) Regular patchy distribution of cytochrome oxidase staining in the primary visual cortex of macaque monkey. Nature 292: 762–764
48. Damasio H, Kuljis RO, Yuh W, Van Hoesen GW, Ehrhardt J (1991) Magnetic resonance imaging of human intracortical structure in vivo. Cereb Cortex 1: 374–379
49. Kuljis RO (1992) Vibrissaeless mutant rats with a modular representation of innervated sinus hair follicles in the cerebral cortex. Exp Neurol 115: 146–150
50. Kuljis RO, Beech RD, Ross SR, Yeung C-Y (1993) Alzheimer-like diffuse amyloid plaques can be induced in transgenic mice expressing human α1-antichymotrypsin. Soc Neurosci Abstr 19: 1035 (421.11)
51. Wahrle SE, Jiang H, Parsadanian M, et al. (2008) Overexpression of ABCA1 reduces amyloid deposition in the PDAPP mouse model of Alzheimer disease. J Clin Invest 118: 671–682

How Biochemical Pathways for Disease May be Triggered by Early-Life Events

Debomoy K. Lahiri, Bryan Maloney, and Nasser H. Zawia

Abstract Alzheimer's disease (AD) is the most common form of dementia among the elderly and usually appears late in adult life. It is presently uncertain when process of this disease starts and how long these pathobiochemical processes take to develop. Therefore, we address the timing and nature of triggers that lead to AD. To explain the etiology of AD, we propose a "Latent Early-life Associated Regulation" (LEARn) model which postulates latent expression of specific genes triggered at the developmental stage of life. This model integrates both the neuropathological features (e.g., amyloid-loaded plaques and tau-laden tangles) and environmental conditions (e.g., diet, metal exposure, and hormones) associated with AD. In the LEARn model, environmental agents could perturb gene regulation in a long-term fashion, beginning at early developmental stages, but these perturbations would not have pathological results until significantly later in life. The LEARn model operates through the regulatory region (promoter) of the gene, specifically through changes in methylation and oxidation status within the promoter of specific genes. The LEARn model combines genetic and environmental risk factors to explain the etiology of the most common, sporadic, form of AD.

1 Prominent Features of Alzheimer's Disease

Alzheimer's disease (AD) is the most prominent form of dementia among the elderly in Western countries [1], estimated to comprise up to 70% of all dementia in the United States [2], afflicting ~5.2 million individuals in 2008 [3]. Its major

D.K. Lahiri
Department of Psychiatry and of Medical & Molecular Genetics Member,
Stark Neurosciences Research Institute
Indiana University School of Medicine
Institute of Psychiatric Research, 791 Union Drive
Indianapolis, IN-46202, USA
e-mail: dlahiri@iupui.edu

R.B. Maccioni and G. Perry (eds.) *Current Hypotheses and Research Milestones in Alzheimer's Disease*. DOI: 10.1007/978-0-387-87995-6_17,
© Springer Science + Business Media, LLC 2009

symptoms include severe loss of memory, failure of cognition and reasoning, and general deficit of other intellectual abilities. Examining its neuropathology reveals excessive deposition of two major proteinaceous aggregates. They are hyperphosphorylated microtubule-associated protein τ "tangles" and senile plaques formed mostly of the amyloid-β peptide (Aβ). Aβ is derived from the amyloid-β precursor protein (AβPP). Amyloid plaques and τ tangles are the basis for the two currently dominant models in the field, the "amyloid hypothesis" and the "τ hypothesis" [4]. The amyloid hypothesis posits that the neurotoxicity of Aβ dimers and oligomers and/or the damage caused by Aβ plaque aggregation are the primary causes of AD. The τ hypothesis maintains that aggregation of hyperphosphorylated τ leads to neuronal cell death and resulting neuropathology. Aβ aggregation would be a result of cellular damage imposed by τ aggregation. Both of these models are "protein-based", and each provides only a proximal cause, not a complete explanation of disease origin.

2 Risk Factors and Genetics of AD

AD is a disorder of complex etiology, combining environmental, genetic, and epigenetic factors. A minority of the cases of AD can be attributed to autosomal familial AD mutations in the coding sequences of AD-associated genes, such as AβPP and presenilin 1 (PSEN1). Unfortunately, this form of AD does not explain the far more common sporadic late-onset AD (LOAD). The known risk factors for sporadic LOAD include age, limited education, head trauma, dietary cholesterol, the APOEϵ4 genotype, and further associations with proteins in addition to AβPP or PSEN1 such as insulin-degrading enzyme, α2-macroglobulin, and endothelin-converting enzyme 2 [5]. In addition, AD risk has been associated with promoter polymorphisms in the APOE and AβPP genes [5, 6]. Oxidative stress and metals in the brain and inflammatory factors are, likewise, linked to AD [7]. However, of the risk factors mentioned, indeed, of all known risk factors, age itself poses the greatest risk for LOAD. None of the current etiological models has proven sufficient to explain the sporadic nature of AD and the "incomplete" effects of known risk factors.

2.1 The LEARn Model Addresses Fundamental Questions in AD Etiology

Although AD manifests late in adult life, it is not clearly understood when the disease actually starts, nor how long it takes for the associated biochemical events and neuropathological processes to develop the disease. Most work has presumed that AD is a late phenomenon, but it could very well be early/developmental in nature. The factors that trigger the cascade of pathobiochemical processes of the disease remain unclear. Major unresolved questions concern the timing and nature of AD triggering. Any unifying hypothesis for the etiology of AD must take into account not only the neuropathological features but also the multiple environmental factors associated with AD, including dietary imbalance, toxicological exposure, hormonal factors, and inflammation after head injury. To address the interaction of environment and genes, that is, in play with AD, we have recently proposed a "Latent Early-life Associated Regulation" (LEARn) model, which may explain the etiology of AD, and possibly several other neuropsychiatric and developmental disorders [8].

The LEARn model begins with environmental agents (e.g., heavy metals), intrinsic factors (e.g., inflammatory cytokines), and dietary factors (e.g., folate and cholesterol) acting to alter gene regulation in a long-term fashion. This begins at early developmental stages, but these perturbations do not yield pathological results until significantly later in life [9, 8]. One such pathological result would be development of AD-like pathology in aged monkeys after infantile exposure to environmental lead (Pb) [10]. Similar hypotheses were developed in the 1980s by Barker et al. [11]. However, Barker's model is predicated upon low birth weight and rapid childhood weight gain without proposing an underlying molecular mechanism. The LEARn model is based on the regulatory structure common to eukaryotic genes and epigenetic processes operating at certain specific sites within the promoter (regulatory) region of specific genes. Overall, LEARn is consistent with the classic "Barker model" of fetal origins of adult disease. It enlarges upon that model by providing specific biochemical and molecular biological pathways that can be directly tested.

2.2 Biochemical Bases of the LEARn Model

In the LEARn model, the foundation of latent early regulation is epigenetic modification of gene regulatory sequences with delayed, latent, changes in gene expression levels. LEARn does not reject the already-present variation of the genetic substrate, but combines it with long-term response to differences in environmental stresses. Human DNA is most commonly modified by DNA oxidation and methylation. Methylation involves the addition of a methyl group to cytosine residues at CpG dinucleotides. This reaction is catalyzed by DNA methyltransferase (DNMT) enzymes. In the DNA sequence, CpG dinucleotides are often found in clusters called CpG islands. In normal tissues, CpG islands are primarily unmethylated, and the aberrant methylation of CpG islands is most likely related to disease. Changing the methylation status of a gene, for example, hypomethylation in the promoter region leads to elevated gene expression, whereas hypermethylation results in decreased gene expression. Environmental factors, including exposure to metals and dietary factors, may operate by interfering with the methylation of CpG clusters, thus altering affinity with potential transcription factors proteins, such as MeCP2 and SP1. The activity of DNMT is reduced by heavy metal (cadmium) exposure [12]. Furthermore, heavy metals such as Pb are known to induce oxidative stress [13] that modulate DNA methylation during malignant transformation [14]. Furthermore, oxidation of D-guanosine to oxo-8-D-guanosine interferes with the DNA-binding capacity of MeCP to a methylated cytosine [15], producing an "effective demethylation" due to oxidative damage of DNA. Histone acetylation may also function in this process, with differences in acetylation occurring in response to DNA methylation or demethylation [16]. This, of course, leads to the issue of chromatin remodeling. However, it is our contention that changes in DNA oxidation and methylation are the fundamental means through which epigenetic differences arise in reaction to the environment.

2.2.1 Latent and Delayed Modification of APP Gene Expression: The LEARn Model at Work

A specific example of LEARn-type activity involves the response of both rat and monkey AβPP genes to early-life exposure to Pb. When Basha et al. introduced Pb acetate into the drinking water of dams of infant rats, levels of AβPP mRNA rose and then fell back to normal levels when Pb was removed from their diet. However, at the age of 20 months, these early-exposed rats' AβPP expression levels and levels of Aβ peptide increased both beyond control rats and beyond levels found in rats exposed to Pb at 20 months of age (Fig. 1). This occurred even though levels of Pb in the early-exposed rats were the same as for nonexposed control rats by this time [9].

This work was extended to primates when cynomolgus monkeys that had been exposed to lead for the first month of life were sacrificed for brain tissue analysis at the age of 26 years. Comparison of their brains with control (nonexposed) monkeys revealed that AβPP and Aβ levels were elevated, but Pb levels were the same between exposed and nonexposed monkeys (Fig. 2). In addition, exposed monkeys showed greater brain amyloid aggregation than did nonexposed monkeys [17]. Furthermore, to extend the possibility of such a model to humans, it has recently been shown that methylation levels for an individual person can change over lifespan in a significant portion of a human population [18].

A different delayed neurological response has also been demonstrated in rats subjected to postnatal inflammation. The treatment resulted in increased susceptibility to seizure in adulthood [19].

3   AD and the LEARn Model

The AβPP protein, Aβ peptide, and τ protein all appear in healthy individuals. Their presence is not a sign of active or incipient AD. Thus, there must be some form of trigger to the disease that is independent of the simple presence of these proteins. The LEARn model would explain developmental triggering and latent expression of the AβPP gene at pathological levels. Studies of knockout animals have indicated that AβPP has necessary functions, although there is redundancy with other AβPP protein family members. Therefore, what would trigger AβPP and Aβ peptides to be

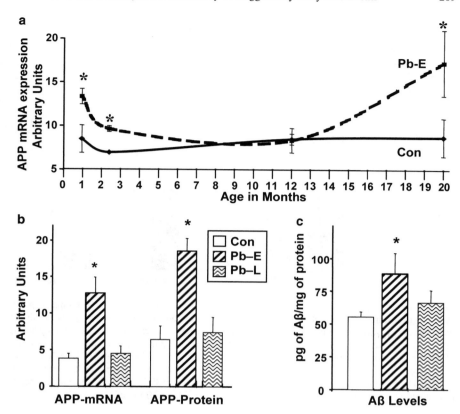

Fig. 1 APP and Aβ levels due to LEARn reaction to post natal dietary Pb in rats. **a** Lifetime changes in rat APP expression in response to early-life exposure to lead (Pb). Rat pups were exposed to dietary Pb through the dams from 1 to 20 days of age (Pb–E rats). Animals were sacrificed at different time points and brain mRNA levels for APP was measured. Results were compared to mRNA levels for the same genes in nonexposed (control) rats. Values marked with an asterisk are significantly different from their corresponding controls ($p < 0.05$). **b** Comparison of end-of-life (20 month) levels of APP mRNA and APP protein, among control, Pb–E, and late-life exposed (Pb–L) rats. **c** Comparison of Aβ peptide levels among Pb–E, control, and Pb–L rats. Levels of APP mRNA and protein and levels of Aβ did not differ between control and Pb–L rats, in contrast to Pb–E rats, which did differ significantly from control rats. Values marked with an asterisk are significantly different from their corresponding controls ($p < 0.05$). This figure was adapted from ref. [9]

overproduced in sporadic cases of AD? The LEARn model proposes that the initial AβPP triggering mechanism activates early in life, at developmental stages. Sites of action would be within the promoter of AβPP and associated genes. The trigger would be maintained through epigenetic means, such as DNA methylation. It is also possible that genes with products protective against AD will have altered methylation patterns due to environmental stress. However, a long-latent condition such as LEARn-induced predisposition to AD is likely to function as a "two-hit" disorder, similarly to those found in currently accepted models of cancer etiology [20]. In the

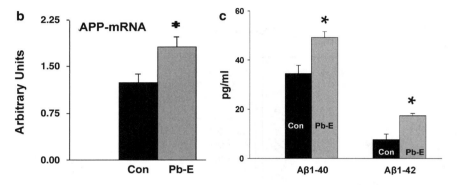

Fig. 2 APP and Aβ levels due to LEARn-type reaction to post natal dietary Pb in monkeys. **a** Pb exposure in cynomolgus control Pb–E monkeys and comparison of end-of-life (23 year) Pb levels between control and Pb–E monkeys. Pb–E cynomolgus monkeys were exposed to Pb in drinking water between 1 and 2 years of age. Control animals were not exposed. Animals were sacrificed after 23 years of life. Pb levels did not differ between control and Pb–E monkeys. **b** Levels of APP1 mRNA and **c** Aβ peptides in Pb–E monkeys versus control monkeys. The mRNA levels in monkey brain were measured by real-time PCR. Aβ peptide levels were measured by ELISA. Levels significantly different from control (p < 0.05) are indicated by "*." This figure was adapted from ref. [10]

case of AD, this second hit would be a broad spectrum of changes in gene expression, especially upregulation of inflammatory factors, that has been shown to be a function of normal aging [21]. It is not to say that these environmental insults "intentionally" target AD-related genes in the brain. Instead, certain genes, by juxtaposition of CpG sites with important active/inactive transcription factor sites, would be particularly vulnerable to the effects of environmental stresses that alter CpG methylation patterns. Testing these changes could, in part, be had by microarray analysis of sporadic and familial AD, if these assays were longitudinal, tracing individual expression profiles across a lifetime and permitting comparison of the same AD and non-AD individuals at multiple life stages.

4   The LEARn Model in Context

The LEARn model is an expression of the larger concepts of epigenetics and the epigenome, specifically adapted to the etiology of sporadic disorders of long latency. The epigenome is the collection of epigenetic markers associated with a specific individual organism's genome [22]. This epigenome, like its underlying genome, has specific epigenotypes, which are generated by modification of DNA methylation or oxidation, by changes in histone acetylation patterns, and by variations in the physical arrangement of chromosomal material [23]. It is the expression of these epigenotypes, whether they be inherited by imprintation or acquired during life as somatic epitypes [24], that specifically contribute to development of sporadic psychobiological diseases.

Of particular interest from a medical standpoint is that the epigenome is inherently less stable than is the genome. The epigenome is subject to epigenetic drift, which is a change in epigenetic markers over time within an individual life span. On a population level, this has been difficult to measure, but epigenetic drift, specifically in the methylation of an individual person's genome, has recently been illustrated in the well-characterized Icelandic sample set and in a cohort of families in Utah [18].

However, the function of LEARn mediation of disease rests upon not only the presence of epigenetic variation but also its location. LEARn-susceptible gene promoters are predicted to have regions of CpG doublets overlapping critical transcription factor binding sites. Modification of the methylation status of these sites is the basis of modification of LEARn-vulnerable gene expression. If such a gene happens to code for a transcription factor or DNA methylation pathway protein, this could result in a LEARn feedback loop, in which local and short-term environmental stress sets up a self-perpetuating cascade that could significantly modify the entire epigenome.

5   Major Implications of the LEARn Model

The LEARn model proposes environmentally induced changes in methylation and oxidative damage as the physical mechanism that perturbs gene expression [25]. The effects of these perturbations would be latent. Lack of an acute response or cessation of acute response would be followed some time later, after a secondary trigger occurred (Fig. 3). They are not always immediately apparent in the same manner found in conventional toxic responses. Thus, apparent reversal of the symptoms of acute exposure to environmental stressors, such as Pb or poor nutrition, need not mean that there will be no long-lasting repercussions of an environmental insult. Under the LEARn model, conventional antitoxicity treatments would be insufficient, as removing the cause does not remove the effect. For example, bans enacted upon Pb in gasoline in previous decades would significantly reduce levels of Pb in urban dwellers. However, the LEARn model would suggest that levels of AD would not be likely to decrease in response until 50–60 years after the bans were enacted, when individuals would begin to reach ages at risk of LOAD but would not have suffered higher levels of childhood Pb exposure. It should be pointed out that the possibility of "latent sequelae" to asymptomatic exposure to Pb was raised over 30 years ago, albeit without proposing a specific mechanism of activity [26].

However, the relatively long term of latency suggests the possibility of biologically based medical remediation. The proposed mechanisms of LEARn, changes in methylation and/or oxidative damage to DNA, do lend toward potential solutions to

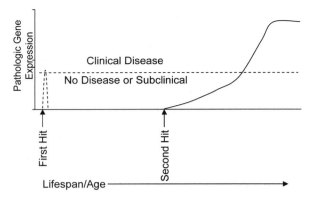

Fig. 3 The LEARn model of disease etiology. This model proposes an epigenetic basis for sporadic disorders such as Alzheimer's disease. Briefly, the sequence of a gene or genes is subject to differences in methylation or oxidation due to environmental stress or insult – the "first hit." These epigenetic changes may have a short-term, nonpathological effect that relatively quickly returns to normal levels of expression. Later, a "second hit," such as typical changes in gene expression profiles brought about by aging has an effect upon genes that have been already epigenetically altered by the first hit. Gene expression then reaches pathological levels

a LEARn-type environmental exposure. For example, fruit juices, such as concentrated apple juice, have been shown to reverse acute oxidative damage and be a useful source of *S*-adenosyl methionine, reversing hypomethylation in mice [27]. Likewise, dietary melatonin supplementation reduced levels of Aβ in mouse cerebral cortex [28]. This suggests investigation of the use of appropriate dietary supplementation early in life, as a prophylactic or treatment measure against possible latent response to environmental insult. In addition, exercise in rats has been shown to modulate the activity of mucosal betaine-homocysteine methyltransferase 2, potentially reducing aberrant methylation [29]. This suggests the possibility that lifestyle habits suspected to protect against AD, such as physical exercise [30], may work through remediation of early-life aberrant DNA methylation. Thus, the LEARn model presented here has far-reaching implications in personal health practices and public policy.

References

1. Hebert LE, Scherr PA, Bienias JL et al. (2003) Alzheimer disease in the US population: prevalence estimates using the 2000 census. Arch Neurol 60:1119–1122
2. Plassman BL, Langa KM, Fisher GG et al. (2007) Prevalence of dementia in the United States: the aging, demographics, and memory study. Neuroepidemiology 29:125–132
3. Alzheimer's Association (2008) Alzheimer's Disease Facts and Figures
4. Sambamurti K, Suram A, Venugopal C et al. (2006) A partial failure of membrane protein turnover may cause Alzheimer's disease: a new hypothesis. Curr Alzheimer Res 3:81–90

5. Lahiri DK, Wavrant De-Vrieze F, Ge Y-W et al. (2005) Characterization of two APP gene promoter polymorphisms that appear to influence risk of late-onset Alzheimer's disease. Neurobiol Aging 26:1329–1341
6. Maloney B, Ge Y-W, Alley GM et al. (2007) Important differences between human and mouse APOE gene promoters with implications for Alzheimer's disease. J Neurochem 103:1237–1257
7. Bellingham SA, Lahiri DK, Maloney B et al. (2004) Copper depletion down-regulates expression of the Alzheimer's disease amyloid-beta precursor protein gene. J Biol Chem 279:20378–20386
8. Lahiri DK, Maloney B, Basha MR et al. (2007) How and when environmental agents and dietary factors affect the course of Alzheimer's disease: the "LEARn" model (Latent Early Associated Regulation) may explain the triggering of AD. Curr Alzheimer Res 4:219–228
9. Basha MR, Wei W, Bakheet SA et al. (2005) The fetal basis of amyloidogenesis: exposure to lead and latent overexpression of amyloid precursor protein and beta-amyloid in the aging brain. J Neurosci 25:823–829
10. Wu J, Basha MR, Brock B et al. (2008) Alzheimer's disease (AD)-like pathology in aged monkeys after infantile exposure to environmental metal lead (Pb): evidence for a developmental origin and environmental link for AD. J Neurosci 28:3–9
11. Barker DJ, Eriksson JG, Forsen T et al. (2002) Fetal origins of adult disease: strength of effects and biological basis. Int J Epidemiol 31:1235–1239
12. Takiguchi M, Achanzar WE, Qu W et al. (2003) Effects of cadmium on DNA-(Cytosine-5) methyltransferase activity and DNA methylation status during cadmium-induced cellular transformation. Exp Cell Res 286:355–365
13. Fowler BA, Whittaker MH, Lipsky M et al. (2004) Oxidative stress induced by lead, cadmium and arsenic mixtures: 30-day, 90-day, and 180-day drinking water studies in rats: an overview. Biometals 17:567–568
14. Campos AC, Molognoni F, Melo FH et al. (2007) Oxidative stress modulates DNA methylation during melanocyte anchorage blockade associated with malignant transformation. Neoplasia 9:1111–1121
15. Valinluck V, Tsai HH, Rogstad DK et al. (2004) Oxidative damage to methyl-CpG sequences inhibits the binding of the methyl-CpG binding domain (MBD) of methyl-CpG binding protein 2 (MeCP2). Nucleic Acids Res 32:4100–4108
16. Dobosy JR and Selker EU (2001) Emerging connections between DNA methylation and histone acetylation. Cell Mol Life Sci 58:721–727
17. Wu J, Basha MR, Brock B et al. (2008) Alzheimer's disease (AD)-like pathology in aged monkeys after infantile exposure to environmental metal lead (Pb): evidence for a developmental origin and environmental link for AD. J Neurosci 28:3–9
18. Bjornsson HT, Sigurdsson MI, Fallin MD et al. (2008) Intra-individual change over time in DNA methylation with familial clustering. JAMA 299:2877–2883
19. Galic MA, Riazi K, Heida JG et al. (2008) Postnatal inflammation increases seizure suscep-tibility in adult rats. J Neurosci 28:6904–6913
20. Knudson AG, Jr. (1971) Mutation and cancer: statistical study of retinoblastoma. Proc Natl Acad Sci USA 68:820–823
21. Lu T, Pan Y, Kao SY et al. (2004) Gene regulation and DNA damage in the ageing human brain. Nature 429:883–891
22. Whitelaw NC and Whitelaw E (2006) How lifetimes shape epigenotype within and across generations. Hum Mol Genet 15:R131–R137
23. van Vliet J, Oates NA and Whitelaw E (2007) Epigenetic mechanisms in the context of complex diseases. Cell Mol Life Sci 64:1531–1538
24. Lahiri DK, and Maloney B (2006) Genes are not our destiny: the somatic epitype bridges between the genotype and the phenotype. Nat Rev Neurosci 7:doi:10.1038/nrn2022-c1
25. Bolin CM, Basha R, Cox D et al. (2006) Exposure to lead and the developmental origin of oxidative DNA damage in the aging brain. Faseb J 20:788–790
26. De la Burde B,and Choat MS (1972) Does asymptomatic lead exposure in children have latent sequelae? J Pediatr 81:1088–1091

27. Chan A and Shea TB (2006) Supplementation with apple juice attenuates presenilin-1 overexpression during dietary and genetically-induced oxidative stress. J Alzheimers Dis 10:353–358
28. Lahiri DK, Chen D, Ge YW et al. (2004) Dietary supplementation with melatonin reduces levels of amyloid beta-peptides in the murine cerebral cortex. J Pineal Res 36:224–231
29. Buehlmeyer K, Doering F, Daniel H et al. (2008) Alteration of gene expression in rat colon mucosa after exercise. Ann Anat 190:71–80
30. Kivipelto M,and Solomon A (2008) Alzheimer's disease – the ways of prevention. J Nutr Health Aging 12:89S–94S

Biomarkers and AD

Strategies for Alzheimer's Disease Diagnosis

Lisbell D. Estrada and Claudio Soto

Abstract Alzheimer's disease (AD) is a devastating degenerative disorder for which there is no cure or effective treatment. In the late stages of the disease when there is evident cognitive impairment, the clinical diagnosis of probable AD is made with around 90% accuracy using clinical, neuropsychological, and imaging methods. Diagnostic sensitivity and specificity in early disease stages are improved by cerebrospinal fluid markers, for example, combined tau and amyloid-β (Aβ) peptides and some plasma markers. At this time, however, validated diagnostic markers for early diagnosis of AD are not available. Simple, accurate, and noninvasive tests for an early detection of AD are urgently required to attempt intervention before substantial and irreversible brain damage occurs. Compelling evidence indicates that deposition of aggregates composed by a misfolded form of Aβ is the central event in the disease pathogenesis. Therefore, an attractive diagnostic strategy is to detect the misfolded form of Aβ in biological fluids. In this chapter, we describe current diagnosis methods, their principles, and their potential strengths and weaknesses. Overall, the available data suggests the feasibility in developing a sensitive diagnostic test for the early detection of AD.

Keywords: Amyloid-β, Alzheimer's disease, oligomers, diagnosis

1 Introduction

Alzheimer's disease (AD) is a complex neurodegenerative condition which has become a major public health problem because of its increasing prevalence, long duration, and high cost of care. It is estimated that more than 25 million people

L.D. Estrada (✉) and C. Soto
Protein Misfolding Disorders Laboratory, George and Cynthia Mitchell Center
for Neurodegenerative diseases, Department of Neurology,
University of Texas Medical Branch, 301 University Boulevard, Galveston TX 77555
e-mail: lisbell.estrada@gmail.com

R.B. Maccioni and G. Perry (eds.) *Current Hypotheses and Research Milestones in Alzheimer's Disease*. DOI: 10.1007/978-0-387-87995-6_18,
© Springer Science + Business Media, LLC 2009

worldwide are affected to some degree by AD. This disorder is characterized by loss of short-term memory, disorientation, and impairment of judgment and reasoning [1–3]. The neuropathological hallmarks found in the brains of AD patients are neuronal loss in regions related to memory and cognition, neurotransmitter depletion, synaptic alteration, and the deposition of abnormal protein aggregates [4, 5]. Neurofibrillary tangles occur intracellularly and are composed of paired helical filaments of hyperphosphorylated tau protein [6]. In the extracellular space, insoluble protein aggregates termed amyloid and composed of the amyloid-β (Aβ) peptide accumulate in the form of senile or neuritic plaques, cerebrovascular amyloid lesions, and diffuse deposits.

2   Current Status of AD Diagnosis

2.1   Neuropsychological Diagnosis

Mental status examinations assess memory, concentration, and other cognitive skills. The most common mental status examination used in the evaluation of AD is the Mini-Mental State Examination (MMSE), a research-based set of questions that provides a score about a person's general level of impairment. The MMSE was designed to evaluate cognitive function in several domains including orientation, registration, attention, memory, language, and visual construction skills [7]. The MMSE is generally a reliable and valid measure of cognitive impairment. However, a number of studies have demonstrated that education, age, and ethnicity have an effect on MMSE scores [8–10]. A study in five different cities in USA found that both age and educational level influenced scores [11]. The MMSE is the most widely used brief screening measure of cognition, but it is not sensitive in detecting mild memory or other subtle cognitive impairments. The Short Test of Mental Status (STMS) was specifically developed for use in dementia assessment and was intended to be more sensitive to problems of learning and mental agility that may be seen in mild cognitive impairment (MCI). Tang-Wai et al. showed that the STMS was slightly more sensitive than the MMSE in discriminating between patients with stable normal cognition and patients with prevalent MCI [12]. Another widely accepted test for AD diagnosis is the Alzheimer's disease Assessment Scale-Cognitive Subscale (ADAS-cog). This test assesses a variety of functions including language ability (speech and comprehension), ability to copy geometric figures, orientation to current time and place and more importantly memory. Still, some weaknesses of the ADAS-cog are the subjective nature of several of the assessments; the fact that it fails to assess several core deficits of AD including attention, information processing and speed of retrieval of information held in memory, the time taken to administer it, and its relative insensitivity (for review see [13]). A possible improvement could be to replace the ADAS-cog with a suitable automated alternative test. However, the major disadvantage of a computerized system is that clinicians are very familiar with the ADAS-cog and are often skeptical about automated tests.

2.2 Electroencephalography

Electroencephalography (EEG) reflects cortical neuronal functioning and remains as an important tool in aiding the diagnosis of AD. The EEG is noninvasive, widely available, low-cost, and can be carried out rapidly. Although a normal EEG is found in many patients with mild AD, the vast majority of patients with moderate-to-severe AD have a pathological EEG profile. The feature of EEG abnormalities in AD is slowing of the rhythms and a decrease in coherence among different brain regions. An increase in theta and delta activities and a decrease in alpha and beta activities are usually observed [14, 15], as well as a reduced coherence of the alpha and beta bands [16–18]. These abnormalities usually correlate with the severity of AD [19, 20]. However, a direct interaction between EEG slowing and AD-associated cognitive dysfunction has not been definitively proven and the progressive EEG slowing is not always present in all AD cases. In addition, a commonly encountered problem in clinical practice during EEG recording is the blanking of the EEG signal due to blinking or movements of the patient's eyes.

2.3 Imaging

The value of structural MRI for AD diagnosis has been recently reviewed [21]. Medial temporal lobe atrophy on MRI has been detected indicating degenerative hippocampal atrophy in old subjects, but is not specific for AD pathology [22]. A PET scan provides both two- and three-dimensional pictures of brain activity by measuring radioactive isotopes injected into the bloodstream. Studies show that an elevated uptake of the PET amyloid ligand (^{11}C)PIB in patients with MCI/mild AD could be indicative of an early AD process [23, 24], even in nondemented individuals [25]. However, at this time, the addition of neuroimaging to the usual diagnostic regimen at AD clinics has been declared not cost-effective given the effectiveness of currently available therapies.

3 Biomarkers

Biomarkers are required to improve the diagnostic sensitivity and specificity and to monitor the biological activity of AD in terms of disease progression (Fig. 1). In view of the advancing scientific knowledge regarding biomarkers for AD, it was proposed to incorporate those biomarkers in revised diagnostic criteria in the future [26, 27]. Biomarkers will initially supplement our more traditional neuropsychological and imaging markers and may progress to useful surrogate measures to the pharmacological action of therapeutic compounds [28, 29].

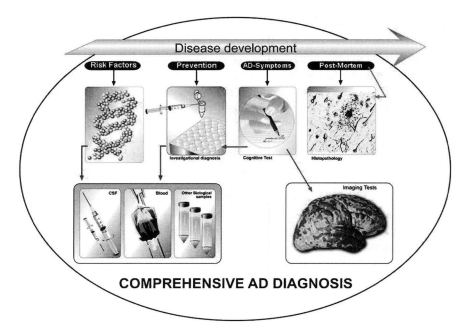

Fig. 1 Current methods of Alzheimer's disease (*AD*) diagnosis such as cognitive tests and brain imaging are based on phenotypic changes, which are the consequence of irreversible brain deterioration. New approaches should aim to diagnose patients during the presymptomatic stages of the disease (*See Color Plates*)

3.1 CSF Biomarkers

The cerebrospinal fluid (CSF) is in direct contact with the extracellular space of the brain, and therefore biochemical changes in the brain affect CSF composition. Aβ42 is the major component of plaques [30]. Therefore, it was postulated as a good candidate for AD diagnosis. Still the first reports of Aβ in CSF as a biomarker for AD were unsatisfactory [31–33]. Recent reports, however, have shown an inverse relation between in vivo amyloid load and CSF levels of Aβ42 in humans [34]. On the other hand, Stefani et al. showed that soluble Aβ42 was not related to the degree of cognitive impairment (2006). In familial cases of AD, extremely high CSF Aβ levels were found in early-onset AD and late-onset AD PSEN1 mutations [35] as well as in children with Down's syndrome [36]. The relationships between plaque density in the brain and CSF Aβ reduction are not completely understood, but they may be due, at least in part, to depletion of the monomeric protein into oligomeric soluble and insoluble forms in the brain and increased Aβ deposition in plaques [37].

Tau presence in CSF was initially proposed as a biomarker for AD by Vandermeeren et al. [38]. It has been reported that antemortem CSF levels of Aβ42, total tau (t-tau), and phosphorylated tau (p-tau) at Thr231 reflect the histopathological changes observed in the brains of AD patients [39, 40]. Also, increase of t-tau CSF levels has been found from early to advanced stages of AD [41]. The CSF tau/Aβ42 ratio has been correlated with cognitive decline in nondemented older adults [42–44] and patients with MCI [45, 46].

The level of p-tau in CSF may reflect the phosphorylation state of tau in the brain. Several reports showed high concentrations of p-tau in CSF of AD patients [47–51]. These studies suggested that CSF p-tau protein correlates with neocortical neurofibrillary pathology in severely demented AD patients and may serve as an in vivo marker of tau pathology in AD [52]. Further studies showed no association of CSF biomarkers (Aβ42, t-tau, and p-tau) with ApoE ε4 or plaque and tangle burden in autopsy-confirmed AD [40, 53]. Still, Apoε genotype and p-tau association with cognitive impairment was reported by Lavados et al. [54]. When they compared AD patients, MCI patients, and normal senile patients, a higher proportion of ε4 allele in both AD patients and the MCI group (with elevated level of impairment) was found. These differences were not statistical significant probably due to the reduced number of observations in the different groups.

Blood inflammatory markers, like C-reactive protein or interleukin-6 (IL-6), are markers for vascular dementia (VaD) [55, 56] or are increased before clinical onset of both AD and VaD [57]. Motta et al. demonstrated that levels of IL-12, IL-16, IL-18, and tumoral growth factor-β1 were higher in mild AD patients, but no significant difference was observed between AD-severe patients and nondemented age-matched subjects [58].

Plasma total amyloid or Aβ42 is increased in cases of familial AD and trisomy 21 [59, 60], but were not consistently related to diagnosis in clinical cross-sectional studies of typical late-onset AD [61–65]. Although all forms of brain Aβ are elevated in AD, the weak correlations of the various brain Aβ measures in AD suggest that they may reflect distinct biochemical and morphological pools of Aβ [66].

Amyloid precursor protein (APP) abnormalities in blood platelets also have been suggested as a biomarker of AD [67, 68], correlating with membrane fluidity and cognitive decline [69]. Recently, a novel gene/protein – ALZAS (Alzheimer-associated protein), has been discovered on chromosome 21 within the APP region. ELISA studies of plasma detected highest titers of ALZAS in patients with MCI (presymptomatic AD). Finally, recent proteomic studies led to the discovery of various plasma signaling proteins altered in AD samples, which may allow the development of novel blood tests for AD [70, 71].

4   A Novel Approach for Biochemical Diagnosis of AD

Simple, noninvasive tests for an early detection of degenerative dementia by use of biomarkers are highly needed. However, up to the present, no accurate extracerebral diagnostic markers for the early diagnosis of AD are available. A sensitive and quantitative test will enable early diagnosis, hence improving drug benefits by allowing the treatment to begin earlier. Although the etiology of AD is not yet completely clear, convincing data suggest that the misfolding, aggregation, and

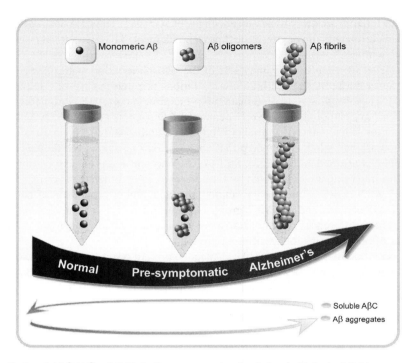

Fig. 2 Amyloid-β (*Aβ*) misfolded oligomers may be circulating in biological fluids many years or even decades before the clinical symptoms of Alzheimer's disease appear. Moreover, their concentration might be indicative of the disease progression. One promising approach may be the specific detection of misfolded Aβ oligomers in biological fluids (*See Color Plates*)

brain deposition of Aβ is the triggering factor of the pathology. Therefore, this misfolding event may be indicative of the initial steps in the disease. Compelling evidence supports that the protein misfolding and aggregation process follows a seeding-nucleation mechanism. In this model, the limiting step is the formation of small oligomeric intermediates that act as seeds to catalyze the polymerization process. It is likely that these misfolded structures are circulating in biological fluids long before the clinical manifestation of the disease. Thus, detection of these oligomeric structures in biological fluids may be the basis to design a diagnostic test for AD and other protein misfolding disorders (Fig. 2). Moreover, convincing evidence support the important role of oligomers in the early pathophysiology of AD, in particular in interfering with synaptic plasticity [72–74].

The challenge is how to detect the very minute amount of these structures and how to distinguish them from the more abundant normal soluble forms. It could be possible to use the functional property of seeds to catalyze the polymerization of a soluble protein as a way to measure the presence of misfolded seeds in biological fluids. With the aim to detect minute quantities of misfolded aggregates, we have developed a technology, termed protein misfolding cyclic amplification, which was first applied to detect misfolded prion protein in the blood of sick animals [75] and has recently been expanded to detect minute quantities of misfolded Aβ oligomers. The method consists of cyclic amplification of seeds combined with the ultrasensitive detection of Aβ aggregates.

5 Concluding Remarks

Today, the only definite way to diagnose AD is to find out whether there are plaques and tangles in brain tissue. To examine brain tissue, however, an autopsy must be performed. Therefore, a diagnosis of AD while a person is still alive constitutes just a possible or probable diagnosis. Several tools are currently used to diagnose "probable" AD, including neuropsychological tests, brain imaging, and biochemical tests. An early, accurate diagnosis of AD helps patients and their families plan for the future. It gives them time to discuss care options while the patient can still take part in making decisions. Moreover, early diagnosis offers the best chance to treat the symptoms of the disease.

Acknowledgments We are very grateful to Marcela Estrada-Escobar for the graphic-designing work. Earlier drafts of this chapter were read by Dr Kristi Green and Zane Martin. All their comments were much appreciated. This work was supported, in part, by CART and Mitchell Foundation.

References

1. Khachaturian ZS (1985) Diagnosis of Alzheimer's disease. Arch Neurol 42: 1097–1105
2. Hamos JE, DeGennaro LJ, Drachman DA (1989) Synaptic loss in Alzheimer's disease and other dementias. Neurology 39: 355–361
3. Wurtman RJ (1985) Alzheimer's disease. Sci Am 252: 62–64
4. Katzman R, Jackson JE (1991) Alzheimer disease: basic and clinical advances. J Am Geriatr Soc 39: 516–525
5. Selkoe DJ (1997) Alzheimer's disease: genotypes, phenotypes, and treatments. Science 275: 630–631
6. Grundke-Iqbal I, Iqbal K, Quinlan M, Tung YC, Zaidi MS, Wisniewski HM (1986) Microtubule-associated protein tau A component of Alzheimer paired helical filaments. J Biol Chem 261: 6084–6089
7. Folstein MF, Folstein SE, McHugh PR (1975) Mini-mental state – practical method for grading cognitive state of patients for clinician. J Psychiatric Res 12: 189–198
8. Blesa R, Pujol M, Aguilar M et al. (2001) Clinical validity of the 'mini-mental state' for Spanish speaking communities. Neuropsychologia 39: 1150–1157
9. Rosselli MC, Ardila AC, Moreno SC et al. (2000) Cognitive decline in patients with familial Alzheimer's disease associated with E280a presenilin-1 mutation: a longitudinal study. J Clin Exp Neuropsychol 22: 483–495
10. Launer LJ, Dinkgreve MAHM, Jonker C, Hooijer C, Lindeboom J (1993) Are age and education independent correlates of the Mini-Mental State Exam performance of community-dwelling elderly. J Gerontol 48: 271–277

11. Crum RM, Anthony JC, Bassett SS, Folstein MF (1993) Population-based norms for the Mini-Mental State Examination by age and educational-level. JAMA 269: 2386–2391
12. Tang-Wai DF, Knopman DS, Geda YE et al. (2003) Comparison of the short test of mental status and the Mini-Mental State Examination in mild cognitive impairment. Arch Neurol 60: 1777–1781
13. Wesnes KA (2008) Assessing change in cognitive function in dementia: the relative utilities of the Alzheimer's disease assessment scale – Cognitive subscale and the cognitive drug research system. Neurodegeneration Dis 5: 261–263
14. Brenner RP, Ulrich RF, Spiker DG et al. (1986) Computerized EEG spectral analysis in elderly normal, demented and depressed subjects. Electroencephalogr Clin Neurophysiol 64: 483–492
15. Coben LA, Danziger WL, Berg L (1983) Frequency analysis of the resting awake EEG in mild senile dementia of Alzheimer type. Electroencephalogr Clin Neurophysiol 55: 372–380
16. Leuchter AF, Dunkin JJ, Lufkin RB, Anzai Y, Cook IA, Newton TF (1994) Effect of white matter disease on functional connections in the aging brain. J Neurol Neurosurg Psychiatry 57: 1347–1354
17. Leuchter AF, Spar JE, Walter DO, Weiner H (1987) Electroencephalographic spectra and coherence in the diagnosis of Alzheimer's-type and multi-infarct dementia. A pilot study. Arch Gen Psychiatry 44: 993–998
18. Locatelli T, Cursi M, Liberati D, Franceschi M, Comi G (1998) EEG coherence in Alzheimer's disease. Electroencephalogr Clin Neurophysiol 106: 229–237
19. Hughes JR, Shanmugham S, Wetzel LC, Bellur S, Hughes CA (1989) The relationship between EEG changes and cognitive functions in dementia: a study in a VA population. Clin Electroencephalogr 20: 77–85
20. Kowalski JW, Gawel M, Pfeffer A, Barcikowska M (2001) The diagnostic value of EEG in Alzheimer disease: correlation with the severity of mental impairment. J Clin Neurophysiol 18: 570–575
21. Vemuri P, Gunter JL, Senjem ML et al. (2008) Alzheimer's disease diagnosis in individual subjects using structural MR images: validation studies. Neuroimage 39: 1186–1197
22. Barkhof F, Polvikoski TM, Van Straaten ECW et al. (2007) The significance of medial temporal lobe atrophy. Neurology 69: 1521–1527
23. Kemppainen NM, Aalto S, Wilson IA et al. (2007) PET amyloid ligand [C-11]PIB uptake is increased in mild cognitive impairment. Neurology 68: 1603–1606
24. Rowe CC, Ng S, Ackermann U et al. (2007) Imaging beta-amyloid burden in aging and dementia. Neurology 68: 1718–1725
25. Pike KE, Savage G, Villemagne VL et al. (2007) β-amyloid imaging and memory in non-demented individuals: evidence for preclinical Alzheimer's disease. Brain 130: 2837–2844
26. Panza F, Capurso C, D,Introno A, Colacieco AM, Capurso A, Solfrizzi V (2007) Heterogeneity of mild cognitive impairment and other predementia syndromes in progression to dementia. Neurobiol Aging 28: 1631–1632
27. Whitwell JL, Weigand S, Przybelski S, et al. (2007) Patterns of atrophy differ between amnestic MCI converters vs non-converters. Neurology 68: A60
28. Blennow K (2005) CSF biomarkers for Alzheimer's disease: use in early diagnosis and evaluation of drug treatment. Exp Rev Mol Diagn 5: 661–672
29. Blennow K, Zetterberg H, Minthon L et al. (2007) Longitudinal stability of CSF biomarkers in Alzheimer's disease. Neurosci Lett 419: 18–22
30. Masters CL, Simms G, Weinman NA, Multhaup G, McDonald BL, Beyreuther K (1985) Amyloid plaque core protein in Alzheimer disease and Down syndrome. Proc Natl Acad Sci USA 82: 4245–4249
31. Van Nostrand WE, Wagner SL, Shankle WR et al. (1992) Decreased levels of soluble amyloid beta-protein precursor in cerebrospinal fluid of live Alzheimer disease patients. Proc Natl Acad Sci USA 89: 2551–2555
32. Farlow M, Ghetti B, Benson MD, Farrow JS, Van Nostrand WE, Wagner SL (1992) Low cerebrospinal-fluid concentrations of soluble amyloid beta-protein precursor in hereditary Alzheimer's disease. Lancet 340: 453–454

33. van Gool WA, Kuiper MA, Walstra GJ, Wolters EC, Bolhuis PA (1995) Concentrations of amyloid beta protein in cerebrospinal fluid of patients with Alzheimer's disease. Ann Neurol 37: 277–279
34. Fagan AM, Mintun MA, Mach RH et al. (2006) Inverse relation between in vivo amyloid imaging load and cerebrospinal fluid A beta(42) in humans. Ann Neurol 59: 512–519
35. Kauwe JSK, Jacquart S, Chakraverty S et al. (2007) Extreme cerebrospinal fluid amyloid beta levels identify family with late-onset Alzheimer's disease presenilin 1 mutation. Ann Neurol 61: 446–453
36. Englund H, Anneren G, Gustafsson J et al. (2007) Increase in beta-amyloid levels in cerebrospinal fluid of children with down syndrome. Dement Geriatr Cogn Disord 24: 369–374
37. Motter R, Vigo-Pelfrey C, Kholodenko D et al. (1995) Reduction of beta-amyloid peptide42 in the cerebrospinal fluid of patients with Alzheimer's disease. Ann Neurol 38: 643–648
38. Vandermeeren M, Mercken M, Vanmechelen E et al. (1993) Detection of tau proteins in normal and Alzheimer's disease cerebrospinal fluid with a sensitive sandwich enzyme-linked immunosorbent assay. J Neurochem 61: 1828–1834
39. Clark CM, Xie S, Chittams J et al. (2003) Cerebrospinal fluid tau and beta-amyloid: how well do these biomarkers reflect autopsy-confirmed dementia diagnoses? Arch Neurol 60: 1696–1702
40. Buerger K, Alafuzoff I, Ewers M et al. (2007) No correlation between CSF tau protein phosphorylated at threonine 181 with neocortical neurofibrillary pathology in Alzheimer's disease. Brain 130: e82
41. Andersson ME, Sjolander A, Andreasen N et al. (2007) Kinesin gene variability may affect tau phosphorylation in early Alzheimer's disease. Int J Mol Med 20: 233–239
42. Fagan AM, Roe CM, Xiong CJ, Mintun MA, Morris JC, Holtzman DM (2007) Cerebrospinal fluid tau/beta-amyloid(42) ratio as a prediction of cognitive decline in nondemented older adults. Arch Neurol 64: 343–349
43. Li G, Sokal I, Quinn JF et al. (2007) CSF tau/A beta(42) ratio for increased risk of mild cognitive impairment – A follow-up study. Neurology 69: 631–639
44. Stomrud E, Hansson O, Blennow K, Minthon L, Londos E (2007) Cerebrospinal fluid biomarkers predict decline in subjective cognitive function over 3 years in healthy elderly. Dement Geriatr Cogn Disord 24: 118–124
45. Hampel H, Teipel SJ, Fuchsberger T et al. (2004) Value of CSF beta-amyloid(1–42) and tau as predictors of Alzheimer's disease in patients with mild cognitive impairment. Mol Psychiatry 9: 705–710
46. Maccioni RB, Lavados M, Guillon M et al. (2006) Anomalously phosphorylated tau and A beta fragments in the CSF correlates with cognitive impairment in MCI subjects. Neurobiol Aging 27: 237–244
47. Blennow K, Wallin A, Agren H, Spenger C, Siegfried J, Vanmechelen E (1995) Tau protein in cerebrospinal fluid: a biochemical marker for axonal degeneration in Alzheimer disease? Mol Chem Neuropathol 26, 231–245
48. Vanmechelen E, Vanderstichele H, Davidsson P et al. (2000) Quantification of tau phosphorylated at threonine 181 in human cerebrospinal fluid: a sandwich ELISA with a synthetic phosphopeptide for standardization. Neurosci Lett 285: 49–52
49. Ishiguro K, Ohno H, Arai H et al. (1999) Phosphorylated tau in human cerebrospinal fluid is a diagnostic marker for Alzheimer's disease. Neurosci Lett 270: 91–94
50. Kohnken R, Buerger K, Zinkowski R et al. (2000) Detection of tau phosphorylated at threonine 231 in cerebrospinal fluid of Alzheimer's disease patients. Neurosci Lett 287: 187–190
51. Hu YY, He SS, Wang X et al. (2002) Levels of nonphosphorylated and phosphorylated tau in cerebrospinal fluid of Alzheimer's disease patients: an ultrasensitive bienzyme-substrate-recycle enzyme-linked immunosorbent assay. Am J Pathol 160: 1269–1278
52. Buerger K, Ewers M, Pirttila T et al. (2006) CSF phosphorylated tau protein correlates with neocortical neurofibrillary pathology in Alzheimer's disease. Brain 129: 3035–3041
53. Engelborghs S, Sleegers K, Cras P et al. (2007) No association of CSF biomarkers with APOE epsilon 4, plaque and tangle burden in definite Alzheimer's disease. Brain 130: 2320–2326
54. Lavados M, Farias G, Rothhammer F et al. (2005) ApoE alleles and tau markers in patients with different levels of cognitive impairment. Arch Med Res 36: 474–479

55. Ravaglia G, Forti P, Maioli F et al. (2007) Blood inflammatory markers and risk of dementia: the conselice study of brain aging. Neurobiol Aging 28: 1810–1820
56. Bibl M, Mollenhauer B, Esselmann H et al. (2008) Cerebrospinal fluid neurochemical phenotypes in vascular dementias: original data and mini-review. Dement Geriatr Cogn Disord 25: 256–265
57. Engelhart MJ, Geerlings MI, Meijer J et al. (2004) Inflammatory proteins in plasma and the risk of dementia: the Rotterdam study. Arch Neurol 61: 668–672
58. Motta M, Imbesi R, Di Rosa M, Stivala F, Malaguarnera L (2007) Altered plasma cytokine levels in Alzheimer's disease: correlation with the disease progression. Immunol Lett 114: 46–51
59. Kosaka T, Imagawa M, Seki K et al. (1997) The beta APP717 Alzheimer mutation increases the percentage of plasma amyloid-beta protein ending at A beta42(43). Neurology 48: 741–745
60. Schupf N, Patel B, Silverman W, et al. (2001) Elevated plasma amyloid beta-peptide 1–42 and onset of dementia in adults with Down syndrome. Neurosci Lett 301: 199–203
61. Tamaoka A, Fukushima T, Sawamura N et al. (1996) Amyloid beta protein in plasma from patients with sporadic Alzheimer's disease. J Neurol Sci 141: 65–68
62. Scheuner D, Eckman C, Jensen M et al. (1996) Secreted amyloid beta-protein similar to that in the senile plaques of Alzheimer's disease is increased in vivo by the presenilin 1 and 2 and APP mutations linked to familial Alzheimer's disease. Nat Med 2: 864–870
63. Vanderstichele H, Van Kerschaver E, Hesse C et al. (2000) Standardization of measurement of beta-amyloid(1–42) in cerebrospinal fluid and plasma. Amyloid 7: 245–258
64. Mehta PD, Pirttila T, Mehta SP, Sersen EA, Aisen PS, Wisniewski HM (2000) Plasma and cerebrospinal fluid levels of amyloid beta proteins 1–40 and 1–42 in Alzheimer disease. Arch Neurol 57, 100–105
65. Assini A, Cammarata S, Vitali A et al. (2004) Plasma levels of amyloid beta-protein 42 are increased in women with mild cognitive impairment. Neurology 63: 828–831
66. Ingelsson M, Fukumoto H, Newell KL et al. (2004) Early Abeta accumulation and progressive synaptic loss, gliosis, and tangle formation in AD brain. Neurology 62: 925–931
67. Borroni B, Colciaghi F, Caltagirone C et al. (2003) Platelet amyloid precursor protein abnormalities in mild cognitive impairment predict conversion to dementia of Alzheimer type: a 2-year follow-up study. Arch Neurol 60: 1740–1744
68. Tang K, Hynan LS, Baskin F, Rosenberg RN (2006) Platelet amyloid precursor protein processing: a bio-marker for Alzheimer's disease. J Neurol Sci 240: 53–58
69. Zainaghi IA, Forlenza OV, Gattaz WF (2007) Abnormal APP processing in platelets of patients with Alzheimer's disease: correlations with membrane fluidity and cognitive decline. Psychopharmacology 192: 547–553
70. German DC, Gurnani P, Nandi A et al. (2007) Serum biomarkers for Alzheimer's disease: proteomic discovery. Biomed Pharmacother 61: 383–389
71. Ray S, Britschgi M, Herbert C et al. (2007) Classification and prediction of clinical Alzheimer's diagnosis based on plasma signaling proteins. Nat Med 13: 1359–1362
72. Lambert MP, Barlow AK, Chromy BA et al. (1998) Diffusible, nonfibrillar ligands derived from Abeta1-42 are potent central nervous system neurotoxins. Proc Natl Acad Sci USA 95: 6448–6453
73. Kayed R, Head E, Thompson JL et al. (2003) Common structure of soluble amyloid oligomers implies common mechanism of pathogenesis. Science 300: 486–489
74. Lacor PN, Buniel MC, Furlow PW et al. (2007) A beta oligomer-induced aberrations in synapse composition, shape, and density provide a molecular basis for loss of connectivity in Alzheimer's disease. J Neurosci 27: 796–807
75. Castilla J, Saa P, Soto C (2005) Detection of prions in blood. Nat Med 11: 982–985

Cognitive Neurology in AD

The Diagnosis of Dementia in Subjects with Heterogeneous Educational Levels

Maria Teresa Carthery-Goulart, Paulo Caramelli, and Ricardo Nitrini

Abstract Performance in cognitive tests is influenced by age, education and culture. Many batteries for screening and diagnosing dementia were developed in countries where low education and illiteracy are uncommon and therefore they may not be suitable for evaluation of populations with heterogeneous educational backgrounds. In this chapter, we discuss the influences of education in the diagnosis of dementia and present the results of studies with the Brief Cognitive Screening Battery (BCSB) or Brief Cognitive Battery unbiased by education (BCB-Edu) developed by Nitrini et al. (1994). The BCB-Edu has been employed for the diagnosis of dementia and Alzheimer's disease (AD) in subjects with heterogeneous educational background (illiterate, low/medium/high levels of education). The delayed recall (DR) test of BCB-Edu proved to be more accurate than DR of Consortium to Establish a Registry for Alzheimer's Disease (CERAD) battery for the diagnosis of dementia in illiterate subjects [sensitivity = 93.3%, specificity = 95.7%, area under receiver operator characteristic curves (AUC-ROC) = 0.975]. DR of BCB-Edu also showed high sensitivity (82.2%) and specificity (90.4%) for the diagnosis of AD in patients with medium/high levels of education. A mathematical model including the results of core tests is also suggested to be employed in clinical practice. The BCB-Edu is a simple and useful tool for diagnosis and screening of dementia, particularly AD, in populations with heterogeneously educated individuals.

1 Introduction

The prevalence of dementia and particularly Alzheimer's disease (AD) will increase dramatically worldwide in the next decades due to the ageing of the global population. In absolute numbers, the elderly population is predicted to grow much faster in least

M.T. Carthery-Goulart (✉), P. Caramelli, and R. Nitrini
Behavioral and Cognitive Neurology Unit, Department of Neurology
University of São Paulo School of Medicine, São Paulo, Brazil
e-mail: carthery@uol.com.br

R.B. Maccioni and G. Perry (eds.) *Current Hypotheses and Research Milestones in Alzheimer's Disease.* DOI: 10.1007/978-0-387-87995-6_19,
© Springer Science + Business Media, LLC 2009

developed and developing countries [1], in which high rates of illiteracy or few years of schooling are common among the elderly [2]. This low educational status is often associated with higher prevalence of dementia among different populations [3, 4]. Among the hypotheses that have been proposed to explain this association, some suggest that education increases brain reserve [5] or is an indicator of mental activity throughout life (cognitive reserve) [6]. However, further research is necessary to establish the possible mechanisms underlying this association and the role of possible confounders such as occupational attainment [7].

Another important problem is functional illiteracy which refers to the "Inability to use printed and written information to function in society, to achieve one's goals and to develop one's knowledge and potential" [8]. According to this definition, around half of the adult world population is considered "illiterate" [9]. In a study conducted in São Paulo in which 312 healthy individuals were evaluated with the Brazilian version of the Test of Functional Health Literacy in Adults (S-TOFHLA), we found that almost 40% of the elderly were considered functional illiterates and another 13% had marginal reading skills. The proportion of these individuals was higher in low-educated groups (Carthery-Goulart et al., submitted for publication). This finding has important implications for epidemiological and clinical studies undertaken with the elderly and highlights the challenges that low-educated individuals present for dementia screening.

Accurate assessment of cognitive functioning is a crucial component in the identification of dementia. Objective evidence of impairment in episodic memory is a core diagnostic criterion for probable AD and reflects the early pathological involvement of medial temporal structures [10]. Neuropsychological tests and particularly memory tests may be influenced by several factors such as age, gender, ethnic background, culture and education [11–14], consequently generating a bias for the diagnosis of dementia. Educational and cultural biases are being gradually more perceived as a problem even in the developed world because of the presence of less-educated and ethnic minorities [15].

The growing prevalence of AD and the emergent treatments for this disease underline the need for reliable and easily administered screening instruments. Some brief cognitive screening batteries have been proposed for this purpose but most of those instruments were developed and validated in countries where low education and illiteracy are uncommon so they produce false positives when applied to populations with lower levels of education. When it happens, those individuals are referred for further assessment which is distressing for them and their families and wasteful of resources [16].

There is a lack of instruments to measure cognitive function in aged populations with low levels of formal education and high levels of illiteracy [16, 17]. In an effort to address this problem, researchers have adjusted cutoff scores or norms of tests according to completed years of schooling [18, 19]. This procedure is frequently employed but has some limitations because low educational level is also a risk factor for dementia [3]. Adjusting for education decreases the risk of observing a false association between dementia and low test performance but increases the probability of not discovering a real association between dementia and low educational level

[17]. Furthermore, it fails to deal with the lack of acceptability to the older person of a testing process tailored to a different population [16].

Other researchers have proposed the adaptation of existent instruments [20] or have tried to devise new tools that could suit the needs of their population. In this attempt, Nitrini et al. [21] developed a cognitive battery that does not require the ability to read and write and that has been found useful for epidemiological and clinical purposes. This battery of tests has been employed in population-based studies and in busy clinic settings for the diagnosis of dementia and AD and, more recently, for the diagnosis of amnestic mild cognitive impairment (aMCI) in subjects with mixed educational background.

2   The Brief Cognitive Screening Battery or Brief Cognitive Battery-Unbiased by Education

The Brief Cognitive Screening Battery (BCSB), recently named Brief Cognitive Battery unbiased by education (BCB-Edu) [22] consists of seven tests of rapid application and simple interpretation: *identification* and *naming* of objects presented as line drawings, *incidental memory*, *immediate memory*, *learning*, *delayed recall (DR)* and *recognition* associated with a category fluency test and the clock drawing test (CDT) [23]. It is available free of charge from the authors.

First, subjects are asked to name ten simple objects presented as line drawings (shoe, house, comb, key, airplane, bucket, turtle, tree, spoon and book). Two scores are obtained, the number of correct named pictures (*naming*) and the number of correct identified pictures (*identification*). If the subject is able to properly name the pictures, the identification is also considered correct. However, if naming is incorrect this could indicate either failure in identification or failure in naming itself and this differentiation is usually straightforward. When naming is incorrect or not remembered, the examiner provides the correct name of the object. Immediately after naming, the sheet of paper is removed from view and the subject is asked to recall the drawings (*incident memory*). The sheet of paper is then shown again for about 30 s and the patient is explicitly asked to memorize the objects. Recall is requested immediately after presentation (*immediate memory*). This latter procedure is performed once more leading to a score of *learning or acquisition*. After two interference tasks evaluating executive function, semantic memory and constructional praxis which are the category verbal fluency test (animals/min) and the CDT, subjects are asked to recall as many items as they can for no more than a minute (*DR*). As a final procedure a sheet of paper with 20 drawings including the 10 previously presented along with 10 added distractors is displayed and the subject is asked to point or name the ones he had previously seen. The score of this recognition test is then calculated by subtracting the number of wrong from the correct responses.

The BCB-Edu was used to study the impact of illiteracy on the performance in DR tests. Nitrini et al. [24] compared the performance of elderly individuals without dementia from a population-based study [4] in two tests of DR, the one from the battery proposed by the Consortium to Establish a Registry for Alzheimer's Disease (CERAD) and the one from BCB-Edu. As the stimuli of the memory test of CERAD battery consist of ten written words, the procedures were adapted for illiterate subjects and the list of words was read aloud by the examiner to those individuals. Except for that, all procedures and tasks of the two batteries were very similar.

In that study, literate and illiterate nondemented subjects presented a considerable difference in performance in most neuropsychological tests. The only tests that were not affected by illiteracy were the BCB-Edu memory tests (Table 1). Illiterate and literate subjects' performances did not differ in the immediate and DR of items presented as simple drawings, contrary to what was observed in the immediate and DR of the word list from CERAD. We proposed that when items to be recalled are presented as simple drawings, encoding is made easier for the illiterate. In the CERAD battery, the different administration procedures required for illiterates determined a less-favorable encoding condition for this group (only auditory input was used while their literate counterparts used visual and auditory input). Our results corroborated findings of other studies in which tests that favor encoding were proved to be more appropriate for illiterate individuals [11, 12, 17, 25].

In a second study [26], the delayed memory tests of CERAD and BCB-Edu were compared for the diagnoses of dementia and AD in literate and illiterate patients from the same population-based study previously mentioned [4]. Test accuracy for dementia and for AD was tested by calculating the area under Receiver Operator Characteristic curves (AUC-ROC) obtained with CERAD and BCB-Edu-delayed memory tests in literate and illiterate samples. Either for dementia or for AD diagnoses, the DR memory test of BCB-Edu was significantly more accurate than CERAD's in the illiterate samples. For the diagnosis of dementia in illiterates, BCB-Edu obtained a sensitivity of 93.3% and a specificity of 95.7% (AUC-ROC = 0.975). Additionally, the two batteries showed similar accuracy in the literate samples. That study showed that the BCB-Edu is a suitable tool for evaluating DR in populations with high number of illiterate and low-educated individuals, a common feature in developing countries.

Table 1 Performance of illiterate and literate nondemented elderly subjects on cognitive tests (adapted from ref. [24])

Tests	Illiterate ($N = 23$); mean (SD)	Literate ($N = 28$); mean (SD)	p^*
MMSE	21.61 (1.85)	26.82 (1.98)	0.000
Verbal fluency (animals/min)	12.13 (3.66)	14.86 (3.93)	0.010
Boston naming (15 pictures)	9.87 (1.82)	12.11 (2.22)	0.001
Clock drawing	3.00 (1.72)	7.52 (1.93)	0.000
Immediate recall (CERAD)	4.04 (1.06)	5.07 (1.05)	0.001
Delayed recall (CERAD)	3.70 (2.10)	4.96 (2.10)	0.014
Immediate recall (BCB-Edu)	6.61 (1.27)	6.57 (1.23)	0.840
Delayed recall (BCB-Edu)	6.87 (1.52)	7.26 (1.56)	0.458

SD standard deviation, *MMSE* Mini-Mental State Examination, *BCB-Edu* Brief Cognitive Battery unbiased by education, *CERAD* Consortium to Establish a Registry for Alzheimer's Disease

As BCB-Edu was initially designed for diagnosing cognitive impairment in subjects who are illiterate or have low levels of education, it was necessary to test if it could also be appropriate as a screening tool for highly educated subjects. This question was addressed in another study [22] and it was demonstrated that the battery is highly accurate and very sensitive for the diagnosis of mild AD in a sample of medium and highly educated individuals (sensitivity of 82.2% and specificity of 90.4%). In that study, we proposed the use of a mathematical model applicable in clinical practice including the scores obtained for DR, learning and category fluency tests. The model comes into a probability score for AD diagnosis so the higher the score the higher the probability of an AD diagnosis. This model was tested in a different sample (from the one it was generated) and obtained an AUC-ROC of 0.917, indicative of high accuracy for mild AD diagnosis. When low-educated individuals are included in the sample to generate the model, the same three tests are included but years of education is incorporated in the equation due to the influences of schooling in the category verbal fluency test (Caramelli et al., in preparation).

The model has recently been tested for the diagnosis of aMCI showing high sensitivity and specificity in the differentiation between these patients and controls (Nitrini et al., in preparation). On the other hand, the specificity of the model for the differential diagnosis between the aMCI and mild AD was not very high. Our results are in agreement with findings that aMCI patients are closer to AD patients than to controls on memory measures [27].

The battery takes about 8 min to be accomplished and presented high interrater reliability [22].

3   Comments, Recommendations and Future Directions

There is an increasing trend to utilize short cognitive batteries for the diagnosis of dementia in epidemiological studies or in busy clinical settings. However, an ideal dementia screening instrument must contemplate several features. It should be acceptable to older persons, brief, reliable and minimally affected by education, gender and age. It should also be accurate (high sensitivity and specificity) and very sensitive to mild prodromal stages of dementia. Additionally, it should be easy to administer and score and should not require excessive training so it can be used in several settings (specialty clinics, epidemiological studies and primary care). Finally, it should be composed by tests evaluating different cognitive functions and specifically cognitive processes considered central to a diagnosis of dementia, particularly AD, the most prevalent form. Thus, it should include at least a test of episodic memory and an indicator of executive function [16, 28].

The BCB-Edu comprises those aspects and is an appropriate tool to be used in populations with heterogeneous educational backgrounds, due to its similar procedural applicability and validity regardless of educational level. It is a culturally and educationally sensitive screening instrument since it does not have items that test arithmetical ability and does not rely on literacy or other skills related to formal education and that often result in unimpaired people screening positive for dementia [16]. Consequently, it can be considered a culture-fair test because it does not penalize members of a culture for poor performance on items designed for and standardized on members of another [14, 17, 20].

Screening tools that entail the ability to read and write as the Mini-Mental State Examination (MMSE) [29] and even others, highly accurate for the diagnosis of dementia such as the Memory Impairment Screen [30] and the ACE-R [31], require excessive adaptation or adjusted cutoff scores to be used with low-educated individuals and may lack specificity for the diagnosis of dementia in these subjects. The CDT, which is often described as a culture free test [16] and is included in batteries such as the Mini-Cog [32], is strongly influenced by education [22, 24] and was not included among the most discriminative tests of the BCB-Edu.

Considering the number of cognitive functions evaluated the BCB-Edu, it is a relatively brief battery. However, screening tests that can be administered in 5 min or less would seem the most likely to be useable in primary care [28]. Further studies are necessary to establish if the CDT can be excluded from the battery without affecting the DR test. It seems possible because some studies showed no difference in DR when a 2–4 min delay was employed between the encoding and recall phases of a memory test [33].

Moreover, the BCB-Edu includes a memory test that optimizes encoding since it requires naming, a cognitive task that demands attention and semantic proce ssing [30]. Other batteries that evaluate episodic memory after forcing semantic processing through naming such as the 7-minute Screen [34] and the *Prueba Cognitiva de Leganés* [17] have been more recently proposed and are also reported as uninfluenced by education.

The BCB-Edu contains tests of learning, DR and recognition and for that reason can be used to characterize the type of memory impairment (encoding, storage and retrieval). This is very important since a diagnosis of AD requires an objective deficit on memory testing (recall deficit with intrusions). Although delayed memory tests discriminate mild AD from healthy controls with high accuracy, impaired DR is not itself evidence of an AD-related memory disorder. Real deficits in encoding and storage processes that are characteristic of AD must be distinguished from non-AD deficits that can also affect DR such as retrieval deficits that may be present for different reasons in normal aging, depression, frontotemporal dementia or subcortical-frontal dementia [10].

The mathematical model including Learning, Delayed Recall and Category Verbal Fluency showed high accuracy in the differential diagnosis between AD and control subjects. This score may be calculated straightforwardly (a template can be provided by the authors under request) and proved very useful for rapid diagnosis of cognitive impairment in a busy setting.

Further validation is needed in other settings and in longitudinal studies, including intraeducation/interculture comparisons. Additionally, the battery needs to be tested in the differentiation of clinical levels of depression and AD and in the differentiation of AD and other dementing disorders.

For dementia screening in heterogeneously educated populations, we recommend the combination of cognitive scores and informant-based tools, as other groups have suggested [35]. Cognitive tests provide an overview of current cognitive performance while informant-rated screens can offer a longitudinal view of changes in cognition-dependant behaviors over time and are less prone to culture bias [16, 28].

It is important to consider that by identifying individuals with a very high or very low likelihood of dementia, a good screening test may play an important part in the diagnostic process. However, no screening tests should replace a comprehensive clinical and neuropsychological evaluation [30].

In conclusion, the advantages of culture-fair screening methods are not restricted to low-educated individuals or minority groups. Those tests present a less-threatening condition in which older people can completely demonstrate their cognitive capabilities, whatever their language, education and culture. This could determine a better screening for all older people [16].

Acknowledgment Fundação de Amparo à Pesquisa do Estado de São Paulo (FAPESP).

References

1. United Nations. Department of Economical and Social Affairs. Population Division (2007) Word Population Prospects: the 2006 revision. Highlights. New York, United Nations
2. UNESCO Institute for Statistics (2006) Education for all global monitoring report. Statistical tables 2006 http://portal.unesco.org/education/en/files/43366/113100877752 adultyouthliteracy. pdf/2adultyouthliteracy.pdf. Accessed 18 January 2008
3. De Ronchi D, Fratiglioni L, Rucci P et al. (1998) The effect of education on dementia occurrence in an Italian population with middle to high socioeconomic status. *Neurology* 50:1231–1238
4. Herrera E Jr, Caramelli P, Silveira AS et al. (2002) Epidemiologic survey of dementia in a community-dwelling Brazilian population. *Alzheimer Dis Assoc Disord* 16:103–108
5. Katzman R (1993) Education and the prevalence of dementia and Alzheimer's disease. *Neurology* 43:13–20
6. Stern Y, Gurland B, Tatemichi TK, et al. (1994) Influence of education and occupation on the incidence of Alzheimer's disease. *JAMA* 271:1004–1010
7. Bonaiuto S, Rocca WA, Lippi A et al. (1995) Education and occupation as risk factors for dementia: a population-based case-control study. *Neuroepidemiology* 14:101–109
8. Kirsch I, Jungeblut A, Jenkins L et al. (1993) Adult literacy in America: a first look at the results of the National Adult Literacy Survey. Washington, DC: National Center for Education, US Dept of Education
9. Coulombe S, Tremblay JF, Marchand S (2004) International Adult Literacy Survey, Literacy scores human capital and growth across fourteen OECD countries. Statistics Canada: Canada's national statistical agency. http://www.statcan.ca/bsolc/english/bsolc?catno=89-552-M2004011. Accessed 18 January 2008
10. Dubois B, Feldman HH, Jacova C et al. (2007) Research criteria for the diagnosis of Alzheimer's disease: revising the NINCDS-ADRDA criteria. *Lancet Neurol* 6:734–746
11. Ostrosky-Solis F, Ardila A, Rosseli M et al. (1998) Neuropsychological test performance in illiterate subjects. *Arch Clin Neuropsychol* 13:645–660
12. Manly JJ, Jacobs DM, Sano M et al. (1998) Cognitive test performance among nondemented elderly African Americans and whites. *Neurology* 50:1238–1245

13. Roselli M, Ardila A (2003) The impact of culture and education on non-verbal neuropsychological measurements: a critical review. *Brain Cogn* 52:326–333
14. Ardila A (2005) Cultural values underlying psychometric cognitive testing. *Neuropsychol Rev* 15:185–195
15. Iype T, Ajitha BK, Antony P et al. (2006) Usefulness of the Rowland Universal Dementia Assessment Scale in South India. *J Neurol Neurosurg Psychiatry* 77:513–514
16. Parker C, Philp I (2004) Screening for cognitive impairment among older people in black and minority ethnic groups. *Age Ageing* 33:447–452
17. Yébenes MJG, Otero A, Zunzunegui MV et al. (2003) Validation of a short cognitive tool for the screening of dementia in elderly people with low educational level. *Int J Geriatr Psychiatry* 18:925–936
18. Caramelli P, Carthery-Goulart MT, Porto CS et al. (2007) Category fluency as a screening test for Alzheimer disease in illiterate and literate patients. *Alzheimer Dis Assoc Disord* 21:65–67
19. Nitrini R, Caramelli P, Herrera E Jr (2005) Performance in Luria's fist-edge-palm test according to educational level. *Cogn Behav Neurol* 18:211–214
20. Ganguli M, Ratcliff G (1995) A Hindi version of the MMSE: the development of a cognitive screening instrument for a largely illiterate elderly population in India. *Int J Geriatr Psychiatry* 10:367–377
21. Nitrini R, Lef'evre BH, Mathias SC et al. (1994) Neuropsychological tests of simple application for diagnosing dementia. *Arq Neuropsiquiatr* 52:457–465
22. Nitrini R, Caramelli P, Porto CS et al. (2007) Brief cognitive battery in the diagnosis of mild Alzheimer's disease in subjects with medium and high levels of education. *Dementia & Neuropsychologia* 1:32–36
23. Sunderland T, Hill JL, Mellow AM et al. (1989) Clock drawing in Alzheimer's disease. *A novel measure of dementia severity. Am Geriatr Soc* 37:725–729
24. Nitrini R, Caramelli P, Herrera Jr E et al. (2004) Performance of illiterate and literate nondemented elderly subjects in two tests of long-term memory. *J Int Neuropsychol Soc* 10:634–638
25. Zunzunegui MV, Cuadra PG, Béland F et al. (2000) Development of simple cognitive function measures in a community dwelling population of elderly in Spain. *Int J Geriat Psychiatry* 15:130–140
26. Takada LT, Caramelli P, Charchat-Fichman H et al. (2006) Comparison between two tests of delayed recall for the diagnosis of dementia. *Arq NeuroPsiquiatr* 64:35–40
27. Petersen RC, Smith GE, Waring SC et al. (1999) Mild cognitive impairment: clinical characterization and outcome. *Arch Neurol* 56:303–308
28. Lorentz BA, Scanlan JM, Borson S (2002) Brief screening tests for dementia. *Can J Psychiatry* 47:723–733
29. Folstein MF, Folstein SE, McHugh PR (1975) Mini-Mental State: a practical method for grading the cognitive state for the clinician. *J Psychiatr Res* 12:189–198
30. Buschke H, Kuslansky G, Katz M et al. (1999) Screening for dementia with the Memory Impairment Screen. *Neurology* 52:231–238
31. Mioshi E, Dawson K, Mitchell J et al. (2006) The Addenbrooke's Cognitive Examination Revised (ACE-R): a brief cognitive test battery for dementia screening. *Int J Geriatr Psychiatry* 21:1078–1085
32. Borson S, Scanlan J, Brush M et al. (2000) The mini-cog: a cognitive 'vital signs' measure for dementia screening in multi-lingual elderly. *Int J Geriatr Psychiatry* 15:1021–1027
33. Kilada S, Gamaldo A, Grant EA et al. (2005) Brief screening tests for the diagnosis of dementia: comparison with the Mini-Mental State Exam. *Alzheimer Dis Assoc Disord* 19:8–16
34. Solomon PR, Hirschoff A, Kelly B et al. (1998) A 7 minute neurocognitive screening battery highly sensitive to Alzheimer's disease. *Arch Neurol* 55:349–355
35. Prince M, Acosta D, Chiu H et al. (2003) Dementia diagnosis in developing countries: a cross-cultural validation study. *Lancet* 361:909–917

Current Anti-Dementia Drugs: Hypothesis and Clinical Benefits

Patricio Fuentes

Abstract At the moment, we have five formally approved anti-dementia drugs. It is well known that abnormalities in cholinergic neurons are prominent among the pathological changes in the brains of patients with Alzheimer's disease (AD) and that the impact of these abnormalities can be reduced by inhibiting the enzymatic breakdown of acetylcholine using cholinesterase inhibitors. Glutamate is the main excitatory neurotransmitter in the central nervous system, implicated in neural transmission, learning, memory and neural plasticity, and the enhancement of the excitatory action of glutamate may play a role in the pathogenesis of AD. Memantine may prevent excitatory amino acid neurotoxicity without interfering with the actions of glutamate that are necessary for learning and memory. However, these symptomatic treatments offer only modest benefits and disease-modifying therapies are still in development stage. Unfortunately, many other drugs with potential anti-pathogenic effects have not shown significant clinical benefits. In mild cognitive impairment, consistent results were not obtained. Then, although exists many pathogenic hypothesis in AD the current and associate available drugs does not change the natural progression of the disease.

Abbreviations AD: Alzheimer's disease; ChEI: cholinesterase inhibitor; DLB: dementia Lewy bodies; FDA: Food and Drug Administration; MCI: mild cognitive impairment; NMDA: N-methyl-D-aspartate

1 Alzheimer's Disease

The research and development of anti-dementia pharmacological agents have been focused on Alzheimer's disease (AD). At present, only five agents have been approved for AD treatment, some of which empirically have also extended their use to other

P. Fuentes
Cognitive Neurology and Dementia Unit, Neurology Service, Hospital del Salvador,
Geriatrics Section, Medicine Department, Hospital Clínico Universidad de Chile, Chile
e-mail: pfuentes@mi.cl

R.B. Maccioni and G. Perry (eds.) *Current Hypotheses and Research Milestones in Alzheimer's Disease*. DOI: 10.1007/978-0-387-87995-6_20,
© Springer Science + Business Media, LLC 2009

dementia etiologies. In AD, different neurotransmitter system abnormalities appear in the cerebral cortex and the hippocampus. Loss of cells in the nucleus basalis of Meynert in AD results in a presynaptic deficiency in acetylcholine in areas of the brain involved in memory, including the hippocampus, cerebral cortex and amygdala [1]. Acetylcholinesterase inhibitors (AChEIs) – tacrine, donepezil, rivastigmine and galantamine – act by inhibiting acetylcholinesterase. They make possible an increase of acetylcholine in the synaptic cleft and stimulate postsynaptic cholinergics receptors. On the contrary, for years, anatomical and biochemical evidence has existed that suggests there is both pre- and postsynaptic disruption of excitatory amino acids pathways, such as glutamate in AD [2]. Based on these previous facts, the other approved drug is memantine, which acts through partial block of NMDA receptors that would allow the prevention of excess stimulation and calcium entrance to the cell [3]. Both glutamate and acetylcholine are related to learning and memory mechanisms.

Many drugs have been developed to control symptoms and also to neutralize some of the pathogenic mechanisms currently known; however, all of these medications currently approved for use in AD are considered palliative and symptomatic treatments. From the formulation of the cholinergic hypothesis [4], the attenuation of the cholinergic deficit has been a major therapeutic target in AD; therefore, the main pharmacological treatment of this dementia has been the use of AChEIs, as we already indicated, including tacrine, donepezil, rivastigmine and galantamine. Summers, in 1986 [5], was first in communicating a successful experiment in cognitive domain with the AChEI tacrine in a small number of patients with AD treated for 12 months. However, this drug exhibited several disadvantages in its use, as the administration frequency had to be every 6 h and the exclusive hepatic metabolism produced drug interactions and hepatotoxicity, with frequent gastrointestinal adverse effects. Later on, the second generation of AChEIs appeared which in numerous clinical trials and several meta-analysis demonstrated, in a consistent way, statistical effectiveness in the global evaluation of cognition, behaviour and functional abilities. Donepezil, a piperidine-specific and reversible AChEI, produces a non-competitive blockade, is absorbed by the digestive tract, and its plasmatic peak appears to 3 h. It is metabolized in the liver by isoenzymes 2D6 and 3A4 of the cytochrome P450 system, with half-life of elimination that reaches 70 h. Drug interactions can lead to hepatic and kidney failure. 5–10% of patients have adverse effects, generally gastrointestinal, with mild intensity and dose dependent. Donepezil can delay cognitive impairment in patients with mild-to-moderately severe AD for at least 6 months duration [6] and also improves cognition and preserves function in individuals with severe AD who live in nursing homes [7]. The third AChEI approved by the Food and Drug Administration (FDA) for AD treatment was rivastigmine, the only drug of this class that in addition inhibits butyrylcholinesterase. It is a carbamate, pseudo-irreversible, non-competitive, with central action, and acts with selectivity on the hippocampus and cerebral cortex. Unlike donepezil, its metabolization does not require the cytochrome P450, but is excreted conjugated to a sulphate by the renal system. The double inhibition that it produces could add cognitive and pathogenic effects since the butyrylcholinesterase would have importance in the transformation in neuritic plaques [8]. It has been demonstrated that with a dose of 6–12 mg/day, the global cognitive and functional impairment of patients with mild and moderate dementia improves significantly, as does their ability to carry out activities of daily living. A recent innovation with this drug has been the formulation in patches, which quickly allows therapeutic doses to be obtained, and a significant reduction of the gastrointestinal effects [9]. Galantamine, the last AChEI drug approved by FDA, is a tertiary amine and has a dual mechanism of action; besides inhibiting cholinesterases, it also modulates a presynaptic nicotinic receptor that would be in greater neurotransmitter release. It is quickly absorbed after oral administration, metabolized by means of the 2D6 isoenzyme of P450, has a plasmatic half-time of elimination of 5 h and is excreted through the urine. Various studies of up to 6 months demonstrate that galantamine also produces cognitive and functional improvement [10]. Marked improvement in cognitive function is usually seen in only a small percentage of subjects, but slowing of decline or stabilization over time has been observed in several trials with these drugs. Through additional analyses, beneficial effects on behavioural symptoms and activities of daily living have been appreciated. Consequently, it is possible to point out that systematic reviews of the available randomized trials, double-blind and placebo-controlled, support the use of these three AChEIs. Statistically significant but clinically modest benefits have been observed with them in each one of the aspects of the disease, in evaluation periods up to 6 months [11], in reducing the risk of functional decline and maintaining the Mini-State Mental Examination for 12 months [12], Even long-term, open-label extension trials have shown some cognitive benefits for more than 24 months [13]. As has been expressed, these drugs, because of their vagal effect, have variable collateral effects and therefore would not be used in patients who have antecedents of chronic bronchial obstruction, peptic ulcer or conduction cardiac alterations.

So these drugs are not free from problems, since we must consider the contraindications, some relative and others absolute, their adverse events and the high cost. Since few comparative trials exist, no convincing evidence demonstrates that one therapeutic treatment is more effective than another. This same class of agents has been used widely in non-Alzheimer dementia, where cholinergic dysfunction is also presumed, although without approval of regulating institutions. There is also accumulating evidence that AChEIs have secondary non-cholinergic functions including the processing and deposition of amyloid-β (Aβ). For instance, treated patients with diffuse Lewy body disease, an entity where cholinergic hypofunction is important, had significantly less parenchimal Aβ deposition [14].

On the contrary, Aβ also plays an important role in the degenerative process of the disease and increases the vulnerability of cultured cortical neurons to glutamate neurotoxicity. Memantine, an uncompetitive, moderate-affinity, voltage-dependent NMDA antagonist, allows normal glutamatergic neurotransmission but blocks potential excitotoxic stimulation. It is presumed to regulate glutamatergic transmission and thereby prevent neuron dysfunction and death. Its half-life is 60–80 h, it does not interact with the cytochrome P450 system, and it is eliminated unmodified by the urine. It is well tolerated and does not have incompatibility, but in case of renal insufficiency, the dose must be adjusted. Although there is some evidence in clinical trials that memantine is functionally neuroprotective, it does have symptomatic benefits. Its effectiveness has been demonstrated in clinical trials, double-blind, placebo-controlled, in patients with moderate or severe AD, also in 6 months of treatment [15]. In a mouse model of AD, the administration of memantine was associated with a significant decrease of Aβ plaque deposition and significant increase in synaptic density [16]. The rationale for numerous medications that are designed to modify the disease process centres on their potential ability to interfere with the known pathobiological cascades in AD. Numerous agents designed to modify the disease process and to delay the speed of declination are currently in developing and evaluation stage. Most of these new molecules are focused in their potential ability to interfere with the pathological process of the disease, like the amyloid cascade or the tau aggregation. The mechanisms of disease-modifying medications may be either specific, with focus on the specific pathological changes in amyloid or tau, or non-specific, including oxidative stress and inflammation or mitochondrial dysfunction. Use of high-dose vitamin E could cause increased risk of cardiovascular disease and a recent study in mild cognitive impairment (MCI) showed no effect in slowing the conversion to AD [17]. As well, studies with oestrogen have shown that it has no effect on progression of symptomatic AD [18]. *Ginkgo biloba* did not yield positive effects in symptomatic AD [19]. Several anti-inflammatory agents have been tested and found ineffective as a treatment for AD [20]. Finally, to date, there is insufficient evidence to suggest the use of statins for the treatment of patients with AD [21].

2 Mild Cognitive Impairment

Clinically, MCI represents a transition from normal cognition to AD, where more than half progress to dementia within 5 years [22]. Inconsistent evidence exists on the usefulness of preventative drugs for conversion; the normal therapeutic suggestion is the recommendation of a healthy lifestyle, trying to control the risk factors and the comorbidities. Clinical trials have focused mainly on the use of AChEIs but the results have been rather negative. A large multi-centre study evaluated the effects of donepezil and vitamin E, over a 3-year period, revealing that donepezil reduced the risk of AD for the first 12 months. The benefit with this AChEI was greater in those subjects carrying the ApoE ε4 genotype, delaying the AD diagnosis for 24 months; but vitamin E did not influence in the time of dementia emergence [17].

3 Conclusions

In AD, but not in MCI, there is a symptomatic benefit with the current approved agents AChEIs and memantine. Although there is evidence of anti-pathogenic actions, the clinical translations of these effects have not yet been determined. Other agents, related to different suggested mechanisms of disease, have not shown convincing effects.

References

1. Whitehouse PJ, Price DL, Clark JT, et al. (1981) Alzheimer disease: evidence for selective loss of cholinergic neurons in the nucleus basalis. Ann Neurol 10:122–126
2. Greenamyre JT, Young AB (1986) Excitatory amino acids in Alzheimer's disease. Neurobiol Aging 10:593–602
3. Parsons CG, Stoffler A, Danysz W (2007) Memantine: a NMDA receptor antagonist that improves memory by restoration of homeostasis in the glutamatergic system – too little activation is bad, too much is even worse. Neuropharmacology 53:699–723
4. Struble RG, Cork LC, Whitehouse PJ, et al. (1982) Cholinergic innervation in neuritic plaques. Science 216:413–415
5. Summers WK, Majovski LV, Marsh GM, et al. (1986) Oral tetrahydroaminoacridine in long-term treatment of senile dementia, Alzheimer type. N Engl J Med 315:241–1245
6. Takeda A, Loveman E, Cleqq A, et al. (2006) A systematic review of the clinical effectiveness of donepezil, rivastigmine and galantamine on cognition, quality of life and adverse events in Alzheimer's disease. Int J Geriatr Psychiatry 21:9–13
7. Winblad B, Kilander L, Eriksson S, et al. (2006) Donepezil in patients with severe Alzheimer's disease: double-blind, parallel-group, placebo-controlled study. Lancet 367:1057–1065
8. Greig NH, Utsuki T, Ingram DK, et al. (2005) Selective butyrylcholinesterase inhibition elevates brain acetylcholine, augments learning and lowers Alzheimer beta amyloid peptide in rodent. Proc Natl Acad Sci USA 102:17213–17218
9. Winblad B, Grossberg G, Frolich L, et al. (2007) IDEAL: a 6-month, double-blind, placebo-controlled study of the first skin patch for Alzheimer disease. Neurology 69 (Suppl 1):S14–S22
10. Wilkinson D, Murray J (2001) Galantamine: a randomized, double-blind, dose-comparison in patients with Alzheimer's disease. Int J Geriatr Psychiatry 16:852–857
11. Qaseem A, Snow V, Cross T, et al. (2008) Current pharmacologic treatment of dementia: a clinical practice guideline from the American College of the Physicians and the American Academy of Family Physicians. Ann Intern Med 148:370–378
12. Mohs RC, Doody RS, Morris JC, et al. (2001) A 1-year, placebo-controlled preservation of function survival study of donepezil in AD patients. Neurology 57:481–488
13. Grossberg G, Irwin P, Satlin A, et al. (2004) Rivastigmine in Alzheimer's disease: efficacy over two years. Am J Geriatr Psychiatry 12:420–431
14. Ballard C, Chalmers KA, Todd C, et al. (2007) Cholinesterase inhibitors reduce cortical Abeta in dementia with Lewy bodies. Neurology 68:1726–1729
15. Reisberg B, Doody Y, Stofller A, et al. (2003) Memantine in moderate-to-severe Alzheimer's disease. N Engl J Med 348:1333–1341
16. Dong H, Yuede CM, Coughlan C, et al. (2008) Effects of memantine on neuronal structure and conditioned fear in the Tg2676 mouse model of Alzheimer's disease. Neuropsychopharmacology, 2008; 33(13):3226–3236
17. Petersen RC, Thomas RG, Grundman M, et al. (2005) Vitamin E and donepezil for the treatment of mild cognitive impairment. N Engl J Med 352:2379–2388
18. Raap SR, Espeland MA, Shumaker SA, et al. (2003) Effect of estrogen plus progestin on global cognitive function in postmenopausal women. The women's health initiative study: a randomized controlled trial. JAMA 289:2663–2672
19. Schneider LS, DeKosky ST, Farlow MR, et al. (2005) A randomized, double-blind, placebo-controlled trial of two doses of Ginkgo biloba extract in dementia of the Alzheimer's type. Curr Alzheimer Res 2:541–551
20. Aisen PS, Schafer KA, Grundman M, et al. (2003) Effects of rofecoxib or naproxen vs placebo on Alzheimer disease progression. JAMA 289:2819–2826
21. Eckert GP, Wood WG, Muller WE (2005) Statins: drugs for Alzheimer's disease. J Neural Transm 112:1057–1071
22. Gauthier S, Reisber B, Zaudig M, et al. (2006) Mild cognitive impairment. Lancet 367: 1262–1270

Index

Printed in the United States of America